T0215521

Self-Service AI with Power BI Desktop

Machine Learning Insights for Business

Markus Ehrenmueller-Jensen

Apress®

Self-Service AI with Power BI Desktop: Machine Learning Insights for Business

Markus Ehrenmueller-Jensen
CEO, Savory Data, Alkoven, Austria

ISBN-13 (pbk): 978-1-4842-6230-6
https://doi.org/10.1007/978-1-4842-6231-3

ISBN-13 (electronic): 978-1-4842-6231-3

Managing Director, Apress Media LLC: Welmoed Spahr
Acquisitions Editor: Jonathan Gennick
Development Editor: Laura Berendson
Coordinating Editor: Jill Balzano

Cover image designed by Freepik (www.freepik.com)

Distributed to the book trade worldwide by Springer Science+Business Media New York, 233 Spring Street, 6th Floor, New York, NY 10013. Phone 1-800-SPRINGER, fax (201) 348-4505, email orders-ny@springer-sbm. com, or visit www.springeronline.com. Apress Media, LLC is a California LLC and the sole member (owner) is Springer Science+Business Media Finance Inc (SSBM Finance Inc). SSBM Finance Inc is a **Delaware** corporation.

For information on translations, please e-mail booktranslations@springernature.com; for reprint, paperback, or audio rights, please e-mail bookpermissions@springernature.com.

Apress titles may be purchased in bulk for academic, corporate, or promotional use. eBook versions and licenses are also available for most titles. For more information, reference our Print and eBook Bulk Sales web page at http://www.apress.com/bulk-sales.

Any source code or other supplementary material referenced by the author in this book is available to readers on GitHub via the book's product page, located at www.apress.com/9781484262306. For more detailed information, please visit http://www.apress.com/source-code.

Printed on acid-free paper

I dedicate this book to my grandma, Franziska.

I hope that all my readers will be as happy and agile at 93 years old as she is now.

Table of Contents

About the Author

Markus Ehrenmueller-Jensen is a data professional who started his career as a consultant for business intelligence solutions on an IBM AS/400 system before he was introduced to the world of Microsoft's data platform in 2006. He has since built data warehouses and business intelligence solutions for a wide variety of clients. His portfolio includes training and workshops, architectural designs, and the development of data-oriented solutions. In 2018, he founded Savory Data, an independent consultancy company.

Markus is certified as a Microsoft Certified Solution Expert (MCSE) for Data Platform and Business Intelligence, and as a Microsoft Certified Trainer (MCT). He teaches databases, information management, and project management at HTL Leonding, Austria. He co-founded both PASS Austria and PUG Austria and co-organizes the yearly SQL Saturday in Vienna. He has been a Microsoft Most Valuable Professional (MVP) each year since 2017.

About the Technical Reviewer

Rodney Landrum went to school to be a poet and a writer. And then he graduated, so that dream was crushed. He followed another path, which was to become a professional in the fun-filled world of information technology. He has worked as a systems engineer, UNIX and network admin, data analyst, client services director, and finally as a database administrator. The old hankering to put words on paper, while paper still existed, got the best of him, and in 2000 he began writing technical articles, some creative and humorous, some quite the opposite. In 2010, he wrote *SQL Server Tacklebox*, a title his editor disdained, but a book closest to the true creative potential he sought; he still yearns to do a full book without a single screenshot, which he accomplished in 2019 with his first novel, *Chronicles of Shameus*. He currently works from his castle office in Pensacola, FL, as a senior DBA consultant for Ntirety, a division of Hostway/Hosting.

Acknowledgments

I want to thank the team at Apress for guiding me through this book project. A shout-out goes to Jonathan Gennick, who helped me to shape the structure of the book.

Rodney Landrum did a brilliant job in being the technical reviewer. He discovered technical issues and inconsistencies in the samples. His hints made the flow of the explanations much better.

A thank you goes to the attendees of my workshops, trainings, webinars, and conference talks. Their questions and feedback are invaluable for me to learn about obstacles in understanding a tool like Power BI. These constantly taught me how to improve my educational material and found their way into the content of this book.

Introduction

Welcome to your journey of AI-triggered features, smart reports, and applied machine learning models in Power BI Desktop!

Power BI

All features described in this book can be used in the free version of Power BI. This includes Power BI Desktop, which you can download from `www.powerbi.com`, and the Power BI Service, which you can find at `app.powerbi.com`. No *Pro* license or *Premium* capacity is necessary to use the features described in this book. In the last chapter, I will show you how you can reach out to the cloud and use Azure's Cognitive Services and Machine Learning Studio. The examples in this book will work with a free subscription of both services. Depending on the number of rows of your productive data, you may hit the boundaries of the free subscriptions though.

In case you are new to Power BI, I included a brief overview of the moving parts, which comprise the following:

- Power BI Desktop
- Power BI Service
- Power BI Report Server
- Power BI Mobile

Power BI Desktop

Power BI Desktop is the so-called client tool. That's the tool you install on your computer. It offers a whole suite of functionalities, including the following:

- Extract, transform, and load data from a wide variety of data sources (with Power Query)
- Build a data model consisting of tables and relationships

- Do all sorts of calculations

- Make your reports (which consist of tables, charts, and filters)

In this book, we take only a limited view on the tool as we concentrate solely on making (smart) reports, do a few calculations, and learn how Power Query will help us understand and enrich the data. Building the right data model is crucial, but is not within the scope of this book. Before you create your first Power BI report on real-world data, make sure that you have carefully read through the following: https://docs.microsoft.com/en-us/power-bi/transform-model/desktop-relationships-understand.

Installing Power BI Desktop

Before you download and open the example files, please make sure to have a current version of Power BI Desktop installed (that means a version of Power BI Desktop released after the book was published). The format of the Power BI (PBIX) files changes constantly—and most likely you will not be able to open a PBIX file created with a newer version of the product than the one installed on your computer.

You can download Power BI Desktop free of charge at www.powerbi.com. Whether you download and install the EXE/MSI or choose the App Store version does not make any difference concerning the available features. The latter updates automatically, which can be an advantage, as there is no real sense in having an old version of the product running.

Power BI Service

You can publish your reports off-premises to Microsoft's cloud at app.powerbi.com. You need a (free) account to do so, but only with a paid license (Pro or Premium) will you be able to share the report with other people. You can find out more about differences between the free version and the paid licenses here: https://docs.microsoft.com/en-us/power-bi/fundamentals/service-features-license-type.

Power BI Service will be mentioned in Chapters 2 ("The Insights Feature") and 4 ("Drill-down and Composing Hierarchies").

Power BI Report Server

You can publish your reports on-premises to Power BI Report Server, which you install on infrastructure you maintain yourself. If you plan to do so, be sure to not use the "ordinary" Power BI Desktop, but rather a version of Power BI Desktop matching the version of your Power BI Report Server. You can find more information about Power BI Report Server via this link: `https://powerbi.microsoft.com/en-us/report-server`.

Unfortunately, at time of writing most of the features demonstrated in this book were not available for Power BI Report Server. If you are using Power BI Report Server, make sure to look up the supported features at this link: `https://docs.microsoft.com/en-us/power-bi/report-server/compare-report-server-service`.

Power BI Mobile App

Don't confuse this client tool with Power BI Desktop. It enables you to consume reports published on either Power BI Service or Power BI Report Server (see previous sections) and is an alternative to using a web browser. Find out more about the app and the available platforms here: `https://docs.microsoft.com/en-us/power-bi/consumer/mobile/mobile-apps-for-mobile-devices`.

Self-Service BI vs. Enterprise BI

Self-service business intelligence (BI) is the term used to describe that someone with domain knowledge builds reports and analysis without the help of IT. The world's most famous self-service BI tool is Excel, closely followed by Power BI.

The advantage of self-service lies in the domain expert's independence from resources in the organization. It enables someone without formal education in IT to discover insights in the data herself. This is only possible when the tool can execute complex tasks in a simple way. Power BI Desktop is exactly such a tool. You will learn many of those features throughout this book, as we are concentrating on self-service BI features that are supported by smart functionalities.

While self-service frees the domain expert, it can lead to data islands throughout the organization. Different people may end up doing similar things, which is a waste of resources. Those people may end up doing the same things in different ways, which can lead to different numbers for the same measure and to meetings where people discuss which of the sales numbers is the correct one.

Here, Enterprise BI comes into play. The goal is to build a centralized data store that contains cleaned data and all the calculations as the single version of the truth. Usually, this layer is called a data warehouse or a cube or a business intelligence semantic model. Building such a layer calls for data engineers who implement the requirements of the domain experts—taking away both effort and freedom from the domain experts.

Finding the right strategy between self-service and Enterprise BI is a balancing act of building the single version of the truth that supports the domain experts in their freedom of gaining insights through their own analysis. This book concentrates on self-service only. Features described and demonstrated in this book can be applied to a central data store as well as to any other data source and can help in gaining insights into data sources so as to enable building the central data store as the next step.

A New Release Every Few Weeks

Writing a book about a piece of software that comes out with a new release every few weeks is a challenge. Power BI Desktop is released monthly. Power BI Service is updated every week. I did my best to not only show you the details of the most current version at the time of writing, but also give you general take-aways, too, which hopefully will be valid for a longer period. All chapters have references to Microsoft's official documentation at `http://doc.microsoft.com`. If your current version of Power BI Desktop looks different than the screenshot in this book, or if you can't find a described feature, please check first which version of Power BI Desktop you have installed (most likely your version will be newer than the version used creating this book). Check back with Microsoft's documentation to find out how the feature evolved.

Artificial Intelligence and All That Jazz

For the sake of this book, I did not stick to any academic definition of artificial intelligence. Moreover, I threw it into the same bucket as machine learning. Some of the features might even be implemented conventionally. I took the view of an end user who does not care about any differences in definitions as long as the available smart feature helps to get to an insight faster.

Chapter Overview

The first chapters will guide you through features that are available via Power BI's graphical user interface alone. You do not need to write a single line of code (except for questions in natural English in the first chapter).

The second set of chapters will show you how you can enhance the functionalities with the help of code. I will introduce and explain code in DAX, M, R, and Python.

The last chapters will go a step further: We will reach out to Microsoft's cloud service in Azure and enhance data while we load it into Power BI Desktop.

Here is the list of chapters:

1.   Asking Questions in Natural Language

2.   The Insights Feature

3.   Discovering Key Influencers

4.   Drill-Down and Decomposing Hierarchies

5.   Adding Smart Visualizations

6.   Experimenting with Scenarios

7.   Characterizing a Dataset

8.   Creating Columns by Example

9.   Executing R and Python Visualizations

10.   Transforming Data with R and Python

11.   Executing Machine Learning Models in the Azure Cloud

Example Database

All examples from all chapters are based on the same dataset, which I loaded from a relational database named AdventureWorksDW. AdventureWorks is a fictive sports item retailer with sales between years 2010 and 2013 for four product categories spread over the globe. You can download a version of this database for SQL Server here for free: `https://github.com/microsoft/sql-server-samples/tree/master/samples/databases/adventure-works`. Make sure to download AdventureWorksDW (with postfix

DW) and not AdventureWorks (without the postfix DW). And make sure to match the version of SQL Server you have at hand (2012, 2014, 2016, or 2017). There are no versions for SQL Server 2019 or later—use the database version for 2017 instead.

Example Reports

The model consists of the following tables and columns:

- Date: Year, Month Number, Date
- Employee: First Name, Gender, Marital Status
- Product: Dealer Price, Category, Subcategory, Name, List Price, Product Line, Standard Cost
- Promotion: Name
- Reseller Sales: Sales Amount, Unit Price, Unit Price Discount
- Sales Territory: Group, Country, Region

The model contains a separate table that collects all measures:

- Discount Amount
- Freight
- Order Quantity
- Product Standard Cost
- Sales Amount
- Tax Amount
- Total Product Cost
- Unit Price AVG

The "Reseller Sales" table is the table with the most rows, as it contains all the transactions (facts). All measures depend on this table. All other tables contain fewer rows and contain filters, which are usually applied on the "Reseller Sales" table to get insightful reports.

I prepared a different Power BI Desktop file (PBIX) per chapter though, to make it easier for you to follow every single step described in each chapter. If a certain table, column, or measure is not used in a chapter, I hid it. You can always unhide those elements if you feel the need to play around with a more complex model or want to try out other columns.

Most of the example files consist of several report pages. Look carefully at the screenshots in the chapters to find out which report page the text is referring to.

CHAPTER 1

Asking Questions in Natural Language

Communicating with computer systems in one's everyday language is a long-held dream of many people. That's why computer languages like COBOL or SQL were born (back in the twentieth century), which are clearly derived from English. But I don't know of many end users who are fluent in either COBOL or SQL. Power BI's *Question & Answer (Q&A)* functionality comes with a better approach. Imported data is automatically enriched by metadata (information about your data), and you can modify this information. This is much more than a convenient way of creating visualizations in a fast way, as you will see in this chapter.

Q&A Visual

First, I will show you how to create a Q&A visual, and then I will walk you through some examples of how to use the visual and improve its behavior.

How to Create a Q&A Visual

We have basically three ways to create a Q&A visual, as follows:

- Choose *Insert – Q&A* from the ribbon.

- Click on an empty space in your report canvas (to be sure to de-select any object on the report page) and then select *Q&A* from *Visualizations*. The icon looks like a speech bubble with a light bulb in its right lower corner.

- Simply double-click any empty space in your report page. That's the fastest way to start the Q&A visual.

1

M. Ehrenmueller-Jensen, *Self-Service AI with Power BI Desktop*, https://doi.org/10.1007/978-1-4842-6231-3_1

Which of the three ways you use is totally your choice. All three give you the same functionality. I definitely prefer the third option (double-click), as this quickly gives me the search box and I can start typing what I am looking for.

Q&A Visual Applied

Figure 1-1 shows what working with a Q&A visual looks and feels like. You have an input field at hand (*Ask a question about your data*) in which you start typing your question (in order to receive an answer from your data). The visual automatically comes up with suggested questions (listed as buttons below the heading *Try one of these to get started*). When you click on one of those buttons, the text is copied into the input field and the answer is generated. Every time you clear the input field, the suggestions will be shown again.

Figure 1-1. *First impression of Q&A visual*

Please open the file ch01_Q&A.pbix in Power BI Desktop. Let's try an example on our own. Ask for sales amount and the Q&A visual will show you 80.45M in a card visual as an answer (Figure 1-2).

80.45M

Sales Amount

Figure 1-2. *Answer to question "sales amount" is 80.45M*

You can narrow your question down by adding `by category` to the existing `sales amount` in the input field. When you enter your question, Q&A might assist you with suggestions of what kind of category you are looking for, for example. Product category is recognized as categorical data, and Q&A comes up with a bar chart in its answer, as you can see in Figure 1-3.

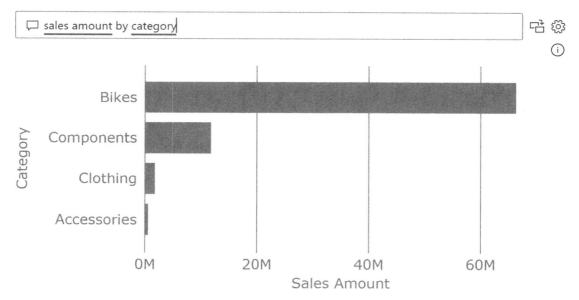

Figure 1-3. *Answer to question "sales amount by category"*

Q&A allows for filtering too. If you add for 2013 (Figure 1-4), Q&A recognizes that you are not asking for a column but rather want to filter on the content of a column, and it will filter the answer for the year 2013 only. In the current view you can only recognize that the filter was applied when watching the numbers on the x-axis (the numbers are now less than half of those in Figure 1-3—even when the relations of the numbers between the categories have hardly changed).

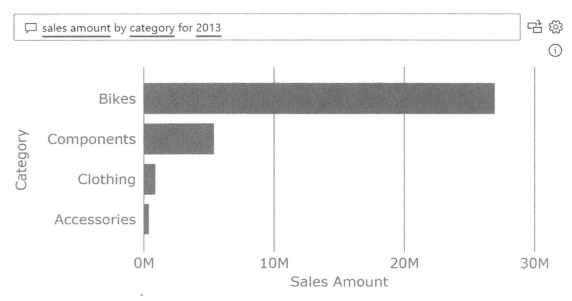

Figure 1-4. *Answer to question "sales amount by category for 2013"*

As soon as we convert the Q&A visual to a "normal" visual (by clicking on the first icon to the right of the input field, which looks like two frames connected by an arrow), you will see that a filter *Year is in 2013* was added to the *Filters* on the right (Figure 1-5). If you don't see the filter pane you can expand it by clicking on the < icon. You can then start changing the visual to your liking, as you now have access to all properties. Unfortunately, converting a Q&A visual to a "normal" visual is one-way. There is no way of converting this visual back to a Q&A visual to change your question. You would have to start over again by creating a new Q&A visual and enter the modified question.

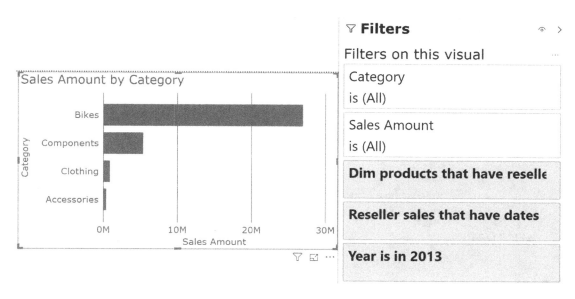

Figure 1-5. *Answer to question is now converted to a "normal" visual, with filters applied*

Now we will create another Q&A visual and ask for `sales amount by sales territory group`. You can see in Figure 1-6 that the answer is a map visual, as `sales territory group` is recognized as geographical information. The map will only be shown if you are currently connected to the internet, as Microsoft's Bing services is used to plot the data on the world map.

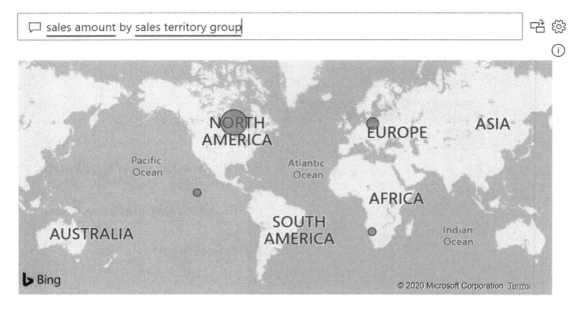

Figure 1-6. *Answer to question "sales amount by sales territory group"*

I don't believe a map is very useful in this case, though. It uses plenty of space on the report page, with little information. And I can expect that my report users know where our sales territories lie on the world map. Let's grip that opportunity and make use of another feature of Q&A: ask to plot the data as bar. You can see the result in Figure 1-7.

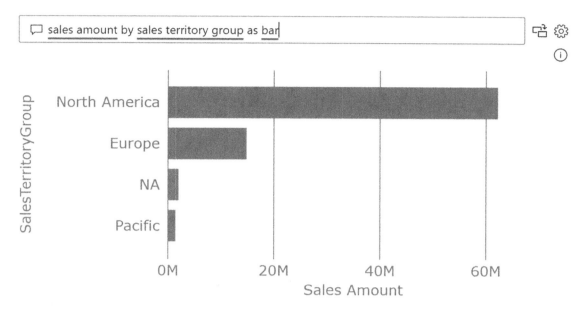

Figure 1-7. *Answer to question "sales amount by sales territory group as bar"*

Most of the standard visuals are supported by Q&A. The documentation says all of them are supported, but I failed in asking for a KPI visual or a slicer. Custom visuals are completely unsupported. That means you can't ask Q&A to show them. But after you have converted the Q&A visual to one of the standard visuals you can of course convert the visual to any visual available.

Watch out, as only terms with a single blue line underneath are understood and considered by Q&A for answering your question (like in the preceding samples). When Q&A does not understand what you are asking for, it puts two red lines underneath the word(s). Terms without any underline are simply ignored (see Figure 1-8).

For example, if you ask for revenue in the sample file, Q&A will not understand the question, as there is no table, column, or measure with such a name available. See the later sections on "Synonyms" and "Training Q&A" for possibilities for teaching Q&A to answer questions about revenue, for instance, with existing sales amount measures.

Hmm, we didn't understand your question. Try fixing double-underlined
terms or ask it another way.

Figure 1-8. *Q&A did not understand what we meant by "revenue"*

Note   I remember that when someone demoed Q&A for me the first time, I silently
said to myself: This is cool for demos—but who will use this feature in real-world
scenarios? It was not very long before I discovered that—if I am familiar with the
data model—I am way faster at creating visuals by just double-clicking the report
canvas and typing in the measures and columns I need, than I am by finding them
by scrolling up and down the field list on the right.

Q&A Button

There is an additional method for opening Q&A: Insert a button onto your report, which
the user can then click to activate the Q&A dialog (not a Q&A visual; for differences, see
Q&A dialog). You find all available buttons under *Insert ➤ Buttons* (Figure 1-9). And
there you can choose Q&A. By default, the button shows a speech bubble icon and no
further text (Figure 1-9).

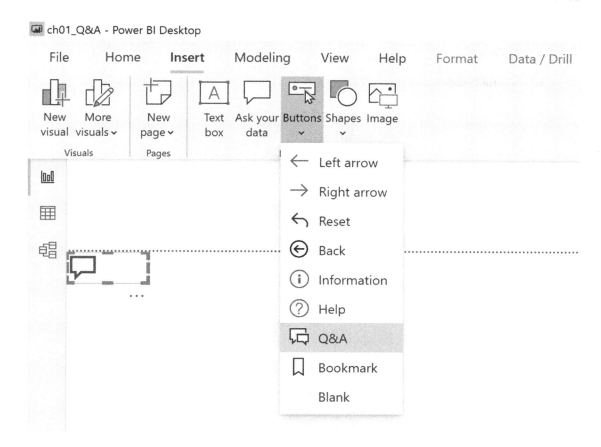

Figure 1-9. *Choose Insert ➤ Buttons to insert a Q&A button*

Under *Visualizations* you have plenty of opportunities to change both the look and feel of the button and its action (Figure 1-10):

- Button Text

- Icon

- Outline

- Fill

- Title

- Background

- Lock aspect ratio

- General

- Border

- Action

- Visual header

Figure 1-10. *Visualizations options for a button*

All the attributes might change when in different states, such as when the button is just visible to the user (*Default state*) or when the user moves the mouse cursor over the button (*On hover*) or when the user actually presses the button (*On press*).

Any space around your entered button text is automatically trimmed. By increasing padding the text moves from the right border toward the center of the frame. You can change the font color, size, and family, and the horizontal and vertical alignment.

For the font color (indeed, for any color Power BI Desktop lets you set) you can choose from:

- Black

- White

- Up to eight theme colors (for more info on themes please refer to `https://docs.microsoft.com/en-us/power-bi/desktop-report-themes`)

- Three lighter (60%, 40%, 20%) and two darker (25%, 50%) versions of the preceding colors. Unfortunately, you can't change those percentages.

- Recently chosen colors

- Custom color, which you can select from a color picker or by entering the hex code of the color

---

Tip   My recommendation is to stick to the first three (black, white, and the main theme colors) as you can change the latter at once by selecting a different theme for your Power BI Desktop file (instead of changing the color format options for every single object in your reports). Be careful with the lighter and darker versions—they might not fit with your CI (corporate identity).

You can change the button's icon to different pre-defined ones (which is not really recommended as it might confuse your users) or remove the button's icon (by setting the icon to "Blank").

Title adds text above your button and offers the same properties as the button text itself, plus you can turn word wrapping on and off (when turned off, too-long texts will be cut off).

The *Outline* definition lets you put a frame around the button. You set the transparency and weight of the frame, and can make the edges round. *Border* puts the frame not only around the button, but the title as well.

Fill lets you change the background color and the transparency of the button itself, while *Background* extends the colored area to the title as well.

Turning *Lock aspect* on or off changes the behavior if you resize the object when you use the handles on the corners of the object (Figure 1-11). It does not change anything for when you use the handles on top, bottom, or the sides of the object. Turning this property on means that the corner handles will resize simultaneously the height and the width of the object. After resizing, the proportion of height to width will stay the same. Turning this property off means that the corner handles will be able to resize the height and width of the object independently. After resizing, the proportion of height to width might have changed.

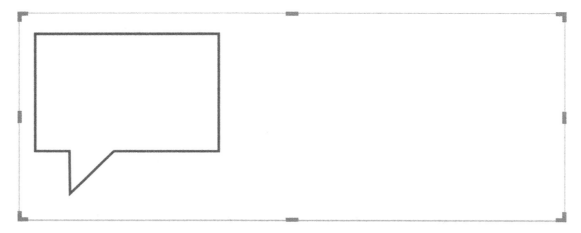

Figure 1-11. *Handles in the four corners act differently depending on option Lock aspect*

General lets you put in exact numbers for the x position and y position of the object and the width and height. Enter a description in *Alt Text* to enable a screen reader to read it when the object is selected.

Action describes what should happen when the user clicks on this button. As we created this button as a Q&A button, the default behavior is to open the Q&A dialog. You can add a tooltip, which is shown as soon as you hover with the mouse cursor over the button.

Visual header shows options that only have an effect after you have published the report to either Power BI Service or Power BI Report Server. Please find more information here: `https://powerbi.microsoft.com/en-us/blog/power-bi-desktop-july-2018-feature-summary/#visualHeader`

Attention   Please remember that within Power BI Desktop you must hold the *Ctrl* key on the keyboard while you click onto the button to activate its action. The reason is simply that just clicking (without holding the *Ctrl* key on the keyboard) on an object in Power BI Desktop is only selecting it. When you have published the report to the Power BI Service or to a Power BI Report Server, the user activates the action via an ordinary click.

Q&A Dialog

The button described in Figure 1-11 does not open a Q&A visualization, but rather a Q&A dialog (Figure 1-12). This dialog shows you suggested questions on the left (which Power BI Desktop is coming up with on its own). You have the same features in the input field (*Ask a question about your data*) with small differences:

- You can see the answer not only to the current questions, but also to previous questions you have asked since the dialog was opened or since the questions were cleared, listed below the input field.

- After asking a question you can narrow it down in a second question after you click *Ask a related question*. A summary of your question is then shown at the bottom of the visual.

- You can clear the list of answers by clicking on *Clear*.

- You can add the current question to the list on the left by clicking *Add this question*. If you leave the dialog window with *Save and close*, these questions are saved with the Power BI Desktop file and replace the standard questions for this dialog. (The questions suggested by default and shown in Figure 1-1 are not influenced, though.)

- You can't add the answer as a visual to your report.

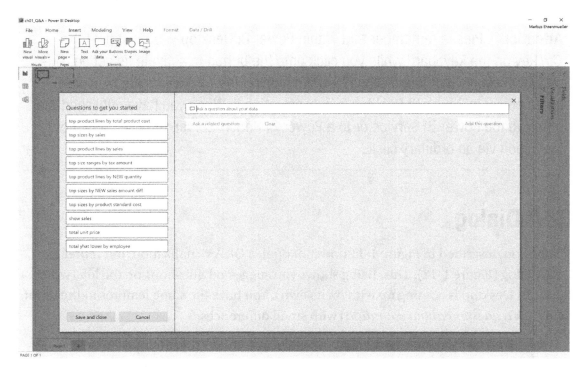

Figure 1-12. *Q&A dialog opened via Q&A button*

When in doubt, I would choose the Q&A visual (created as visual or by double-clicking the report pane) over the Q&A dialog (created via a Q&A button). The Q&A visual has been reworked with the December 2019 release of Power BI Desktop to cover most of the features of the dialog and can be directly embedded into the report page, without the need to open an extra dialog window.

Keywords

We have already seen some possibilities of Q&A in action: getting the value of a measure or a column, filtering on a year, or changing the visual. The first one refers to objects in the data model. The latter two make use of keywords.

The keywords can be categorized in the following way:

- Aggregates (e.g., total, sum, count, average, largest, smallest)

- Comparison and range (e.g., versus, compared to, in, equal, =, between), top x, conjunctions (e.g., and, or, nor), and contractions (e.g., didn't)

- Query commands (e.g., sort, ascending, descending), visual types (e.g., as bar, as table; for me most of the standard visualizations worked)

Beside the keywords you can enter values (concrete values you are looking for, including true, false, and empty), including dates (concrete values in many formats and relative dates, like today, yesterday, previous, x days ago) and times.

You can find a full list of keywords and values in Microsoft's online documentation at `https://docs.microsoft.com/en-us/power-bi/consumer/end-user-q-and-a-tips`. Microsoft does a great job in adding many variations of recognized words. Just try it out (and don't restrict yourself to the preceding list) to find out which phrases work best for you to get the result you are looking for.

Synonyms

Despite the fact that a table, a column, or a measure in a data model can only have one distinct name, in the real world people use different names for the same thing. People in your organization (maybe including yourself) might sometimes refer to the sales amount and at other times to revenue or earnings, but mean the same thing in all three instances. If we have several names for the same thing, those several names are called synonyms.

When you ask Q&A for revenue in the sample Power BI Desktop file, it will put two red lines underneath, signaling that it does not recognize the word (as it can't be found in the data model). We have already seen this in Figure 1-8.

Luckily, Power BI Desktop allows us to enrich our data model with synonyms. The feature is a little bit hidden, though. You have to change to the *model* view (the third icon from the top on the left side of the screen; see Figure 1-13). There, you can either click on a specific element (e.g., a column or measure) in one of the tables or find the element in the field list on the right of the screen. In the *General* section of *Properties* you will find the input field for *Synonyms*.

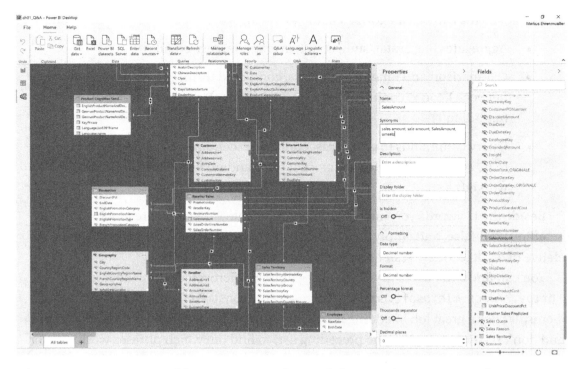

Figure 1-13. *We can add synonyms in the model view of Power BI Desktop*

Power BI Desktop pre-fills this field automatically. It adds its name (e.g., SalesAmount) and variations (e.g., singular `sale amount` and plural `sales amount`). You can add further synonyms by separating the list with a comma. In the example file, I added `umsatz` to the list, which stands for sales amount in German. (You not only learn something about natural language queries in Power BI Desktop in this book, but also extend your own language skills!)

If you add `revenue` to the list (separated with a comma from the existing items in the input field) and change back to the report view, Q&A will be able to find 80M of sales amount as an answer to your question about `revenue`.

Finding out about all synonyms in your organization and then entering them one by one in the Power BI Desktop file can be a challenge. The next two sections will describe further solutions for that problem.

Teach Q&A

Instead of changing to the model view and finding the field where you want to add a synonym, you have another option: If you click on a term that has two red lines underneath, you can opt to teach Q&A the meaning of the term.

To try this out, enter earnings in Q&A and then click on the term. As you can see in Figure 1-14, a bubble will appear, telling you that the folks behind Q&A *aren't sure what you mean*. They ask you to try *another term or add this one*. By clicking on the blue text *Define earnings*, a dialog appears to teach Q&A.

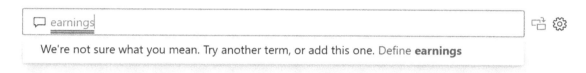

Hmm, we didn't understand your question. Try fixing double-underlined terms or ask it another way.

Figure 1-14. *You can define a term that Q&A did not understand*

A dialog window *Q&A setup* opens, and the section *Teach Q&A* is pre-selected. To teach Q&A what we mean by "earnings," click on the *Submit* button at the right of the input field (Figure 1-15).

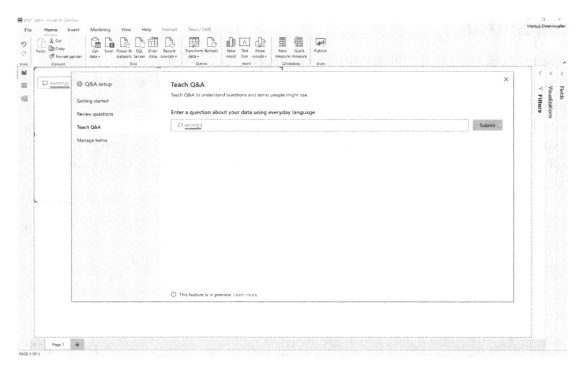

Figure 1-15. *Submit the term Q&A did not understand in the Teach Q&A dialog*

After you click *Submit* (Figure 1-16), you can specify that "earnings" is just another word for "sales amount."

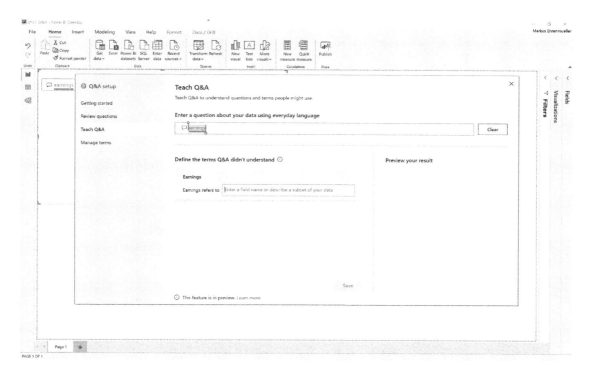

Figure 1-16. *Define the term Q&A did not understand by entering a field name or a Q&A question*

Enter `sales  amount` here and the term "earnings" will be underlined in blue, signaling that Q&A now knows what we were asking for (Figure 1-17).

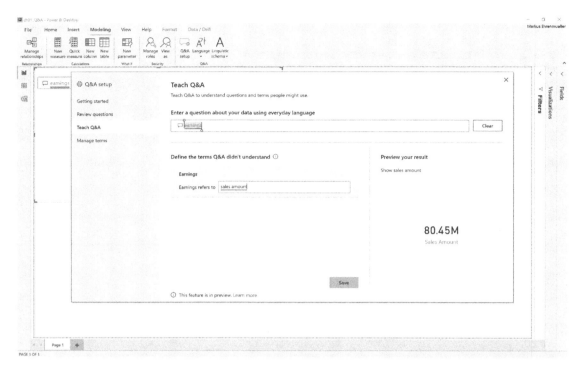

Figure 1-17. *Teach Q&A that "earnings" refers to sales amount*

Depending on the phrasing of your original question, the term definition may contain a more complex phrase with different keywords.

Alternatively, you can start this dialog by clicking on the gear icon, to the right of the input field within the Q&A visual. Then it opens with the section *Getting started* pre-selected (Figure 1-18). This screen contains a description of the three other sections (*Review questions*, *Teach Q&A*, and *Manage terms*) and a link and a video to learn more about Q&A. *Teach Q&A* we have already discussed. Let's look at the two others.

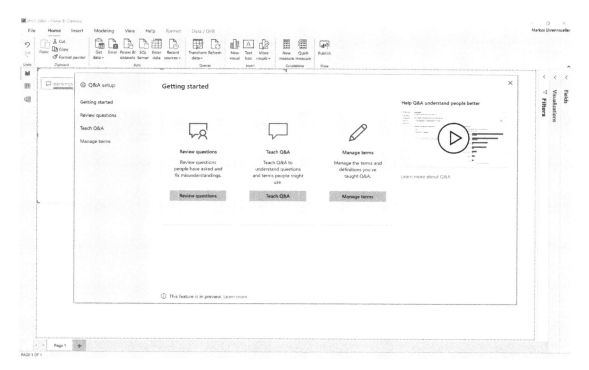

Figure 1-18. *Getting started section of Q&A setup*

In the section *Review questions*, you first have to select one of the datasets you have previously published to Power BI Service before you can get a list of questions that Q&A could not understand. In Figure 1-19 you can see that both revenue and earnings have been asked for in the Power BI Service. Earnings we already taught successfully, but for revenue there is a *Fix needed*. Just click the pen icon to open *Teach Q&A* and tell Q&A that revenue is referring to sales amount as well. As soon as you upload the file to Power BI Service, your colleagues can successfully ask for revenue and earnings as well. Please read https://docs.microsoft.com/en-us/power-bi/desktop-upload-desktop-files for more details of how to publish and share your Power BI file in Power BI Service.

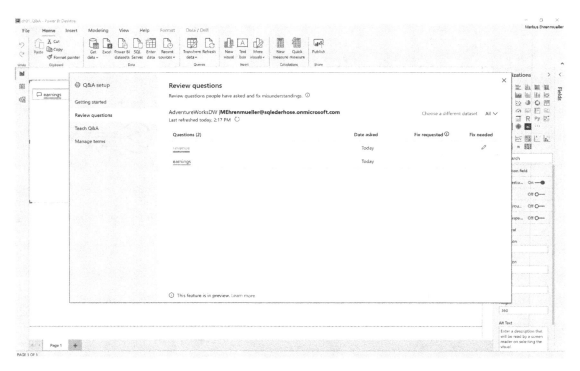

Figure 1-19. *Review questions section of Q&A setup*

The *Manage terms* section reveals all terms we have taught Q&A. So far this is only "earnings," as you can see in Figure 1-20. If the term is not used anymore, you can delete it here. If you want to change the meaning of the term, delete it as well and then enter the new meaning in section *Teach Q&A*.

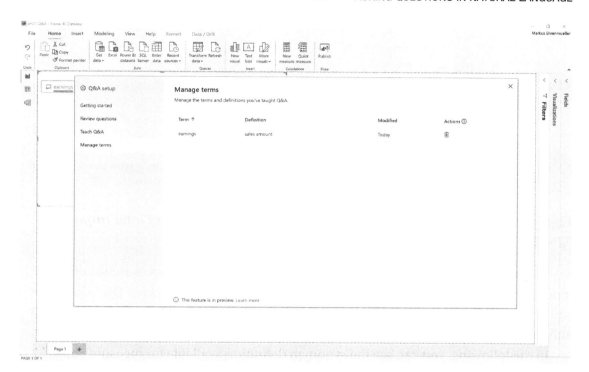

Figure 1-20. *Manage terms section of Q&A setup*

Attention   At the time of writing of this book, Q&A Setup was in preview. I do not recommend using preview features of any kind for analyses that are crucial for you or your organization. Preview features might be unstable and are unsupported, might be changed fundamentally before they make it into the product, and you don't know when or if they will make it into the product. Find updates on this feature here: `https://docs.microsoft.com/en-us/power-bi/natural-language/q-and-a-tooling-teach-q-and-a`.

Linguistic Schema

If you want to bring Q&A to the next level, you can export (*Modeling* ➤ *Linguistic schema* ➤ *Export*), modify, and re-import (*Modeling* ➤ *Linguistic schema* ➤ *Import*) a linguistic schema from your Power BI Desktop file (Figure 1-21).

Figure 1-21. *The existing linguistic model can be exported (and later imported again)*

The schema is a text file (in YAML format) that basically matches elements of the model to synonyms. Those synonyms can have further attributes to help Q&A decide in which cases to use them. This file will contain both all synonyms (from the model view) and taught terms (from *Q&A setup*).

In the following, you can see the synonym definition available for measure *Sales Amount* (which we have seen in the sections "Synonyms" and "Teach Q&A"). Everything but the three lines with - umsatz, - revenue: {State: Deleted}, and - earnings: {LastModified: '2020-02-10T09:14:44.6697036Z'} were automatically generated by Power BI Desktop.

umsatz is the synonym I provided in the sample file. I entered revenue and earnings in *Teach Q&A* and later deleted the first again (hence the state *deleted*).

You can spot in the first line just underneath those three lines that sale quantity was added as a synonym as well (state is *suggested*), but only weighted 0.500000019868215. If there is a synonym *sale quantity* for another column in the model with a bigger weight, this other term will be preferred over *SalesAmount* when we ask Q&A for *sale quantity*. More suggested synonyms, with different weights, were added as well.

```
v_Measures_measure.sales_amount:
  Binding: {Table: _Measures, Measure: Sales Amount}
  State: Generated
  Terms:
  - sales amount
  - sales: {Weight: 0.97}
```

```
- umsatz
- revenue: {State: Deleted}
- earnings: {LastModified: '2020-02-10T09:14:44.6697036Z'}
- sale quantity: {Type: Noun, State: Suggested, Weight:
0.500000019868215}
- sale volume: {Type: Noun, State: Suggested, Weight:
0.49999991986821091}
- sale expanse: {Type: Noun, State: Suggested, Weight:
0.499999869868209}
- sale extent: {Type: Noun, State: Suggested, Weight:
0.499999819868207}
- salary: {Type: Noun, State: Suggested, Weight: 0.49090911041606561}
- paycheck: {Type: Noun, State: Suggested, Weight: 0.49090896314333249}
- remuneration: {Type: Noun, State: Suggested, Weight:
0.49090886496151037}
- retribution: {Type: Noun, State: Suggested, Weight:
0.49090881587059931}
- sale sum: {Type: Noun, State: Suggested, Weight: 0.48500001927216851}
- sale total: {Type: Noun, State: Suggested, Weight:
0.48499997077216667}
```

Key Takeaways

This is what you have learned in this chapter:

- Creating new visuals is way faster by double-clicking on the report pane (to start Q&A) and typing in the measures and columns needed. After converting this into a visual you can then tweak the visual to your needs.

- Q&A offers you not only the ability to select the elements for the visual but also an understanding of natural language that is used to calculate aggregations, apply filters, or choose the type of visual.

- Q&A comes in two tastes: Q&A visual (activated via double-click on the report canvas) and Q&A dialog (activated via added button).

- You can add synonyms to your model to help Q&A better understand the (business) terms the user is using.

- You can train Q&A to recognize more complex terms by providing a description in natural language.

- The most advanced method to help Q&A recognize the daily language of you and your organization is to provide a linguistic scheme in the form of a YAML file.

CHAPTER 2

The Insights Feature

To get answers with Q&A (see previous chapter), you have to come up with your own question. In some cases, the question is not so easy to formulate—it would be much easier to just click on a data point in your visualization and let Power BI find the reason why this data point is different from the others (e.g., why the value for one month is much higher or lower than that for the previous month). The Insights feature does exactly that. Again, this functionality is dependent on metadata available from the data model. An insight can be either converted to a standard visual or pinned to a Power BI Service dashboard and so easily shared with your report.

Explain the Increase

Take the example from Figure 2-1 and watch it closely. It shows a line chart for sales amount by date over a range of three years. While sales have been volatile over those months, there is a clear peak in January 2013. From an analytics point of view, it could be interesting to find out what went differently in January 2013 compared to December 2012. Maybe we can derive a pattern from these two data points and can learn what we could do to replicate this in the future.

© Markus Ehrenmueller-Jensen 2020
M. Ehrenmueller-Jensen, *Self-Service AI with Power BI Desktop*, https://doi.org/10.1007/978-1-4842-6231-3_2

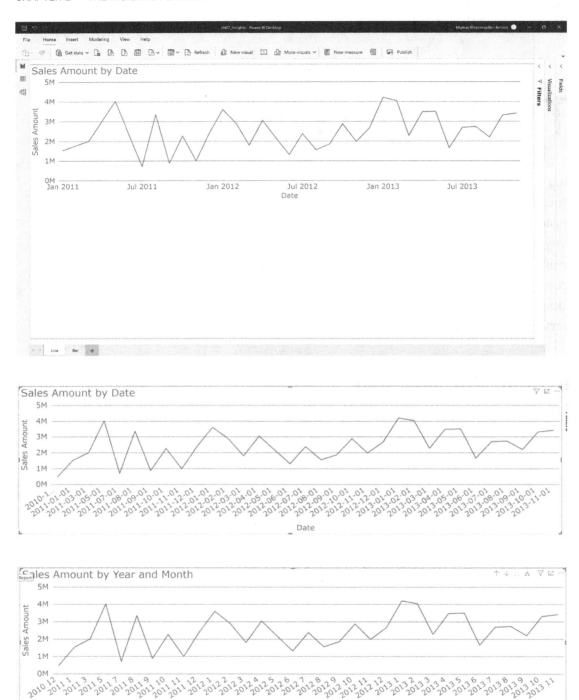

Figure 2-1. *Sales Amount by Date as a line chart*

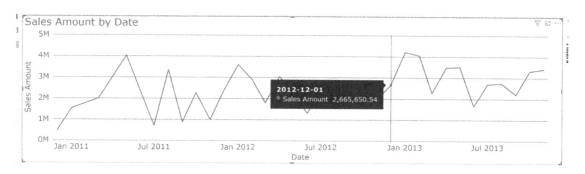

Figure 2.1. *(continued)*

To find out, simply right-click the peak data point in January 2013 and select *Analyze* ➤ *Explain the increase.* Power BI Desktop will take a few seconds to analyze all available information in your data and data model and then suggest explanations that are statistically significant (Figure 2-2).

Attention   Not everything that is statistically significant is practically significant. Weigh the suggestions against your domain knowledge and your common sense.

Even when something has a practical significance it does not tell us directly which of the insights was the cause and which one was the effect. Again, we need domain knowledge and common sense to draw the right conclusions.

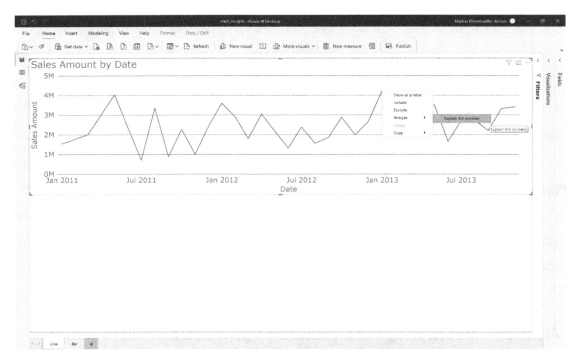

Figure 2-2. *Right-clicking a data point offers to explain the increase*

Power BI Desktop shows the insights in a flyout window (Figure 2-3) with the following heading: *Here's the analysis of the 58.05% increase in Sales Amount between Saturday, 1 December 2020 and Tuesday, 1 January 2013*. It came up with a list of suggestions to explain the increase, which you can see if you scroll down inside the flyout window: Sales Amount by Date and ...

- First Name

- Gender

- Marital Status

- English Promotion Name

- Product Line

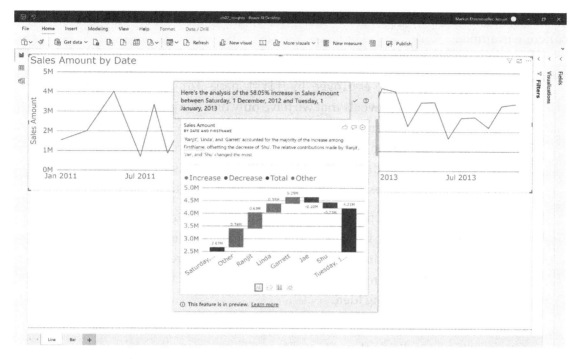

Figure 2-3. *Here's the analysis of the increase*

Each one has a detailed description of the insight in plain English. Let's take a closer look at the first one: Sales Amount by Date First Name.

'Ranjit,' 'Linda,' and 'Garrett' accounted for the majority of the increase among FirstName, offsetting the decrease of 'Shu.' The relative contribution made by 'Ranjit,' 'Jae,' and 'Shu' changed the most.

Below, we see the same information in the form of a waterfall chart. It shows a blue bar on the left representing the sales amount in December 2012 and one at the right representing January 2013. The change between those two points in time is showed as a positive contribution to the increase of sales amount by Ranjit, Linda, and Garrett and a negative one by Jae and Shu. Employees with lesser contributions (positive or negative) are not shown in detail but rather are aggregated under a gray column named *Other*.

This might be a good insight in some cases, but I doubt that it is a good insight in the concrete example:

- The first names of our employees are not unique. A quick check via Q&A with "`count of employee by first name for Ranjit Linda Garrett Jae Shu`" reveals that we have only one Ranjit, one Jae, and one Shu, but two Garretts and four Lindas. The bar in the waterfall chart is adding up the values of those four Lindas—we can't tell which of the Lindas did a good job or if all four did just an average job in increasing the sales amount from December 2012 to January 2013.

- Even if we know exactly which of our employees contributed how much to the increase in sales amount between December 2012 and January 2013, what kind of conclusion can we draw from this kind of information? Hiring more people with the first name Ranjit would probably be the wrong idea.

- With data privacy in mind (e.g., EU-GDPR), we should maybe not do reporting on individual employees.

But why does the Insights feature suggest the first name at all? The reason is simply because the column is available in the data model. And Power BI—agnostic about the meaning—analyzes all available information. To get rid of the suggestion involving first names, we can hide the column in the data model. A hidden column will still be there (and can be used in filter relationships in the model, be shown in reports, or be part of a DAX calculation). It will only disappear from the field list, and the Insights feature (and Q&A as well) will ignore the column.

To hide `FirstName`, find it in the field list on the right. Right-click it and select *Hide* from the context menu, like you can see in Figure 2-4. In case you change your mind later, you can un-hide it again: Right-click any object in the field list and select *View hidden*. All the hidden elements will appear but will be greyed out. You can then right-click any hidden element and de-select *Hide* from the context menu.

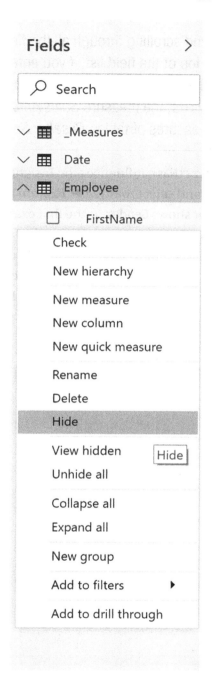

Figure 2-4. *Right-click a column to be able to hide it*

Note   Instead of clicking and scrolling through all the fields in the field list, you can use the search field on top of the field list. If you enter text, the field list is automatically limited to tables, columns, and measures containing the entered text. For example, if you enter `sales`, the measure *Sales Amount* will be listed as well as all of the columns and measures of tables *Reseller Sales* and *Sales Territory*.

Let's see how removing `FirstName` influences the Insight feature. Again, right-click January 2013 in the line chart and select *Analyze* ➤ *Explain* the increase. This time, Insights ignores first name, but shows `Gender` as the first example. We can clearly see that the majority of the increase in sales amount was due to the fact that male employees sold 1.17M more (Figure 2-5). Female employees sold more as well, but only 0.38M. To assess if our male salespersons did a better job than the female ones, we would have to know how the spread between males and females is in first place. If we have three times more male employees than female ones, the spread from 1.17M to 0.38M would not come as any surprise and would be expected on average.

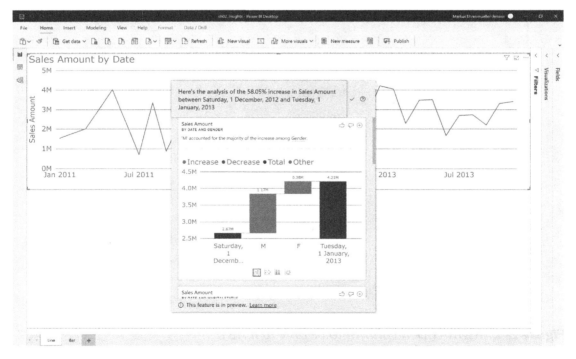

Figure 2-5. *Here's the analysis of the increase, after we hide first name*

We can find out quickly within the Insights flyout window. Below the waterfall chart in Figure 2-5 you can see four icons, standing for four different kinds of visuals available for the current insight. First comes *Waterfall chart*, which is the pre-selected one. Then comes *Scatter chart*, *100% Stacked column chart*, and *Ribbon chart*.

The *Scatter chart* is not very helpful in this case, so we can ignore it. The latter two are interesting, as they show the distribution of gender for the two dates. The first one shows equally high columns (a *100% Stacked column chart*), visualizing the distribution in percentage. The second one shows two stacked columns of different height with the actual values of sales amount by gender. The gender portions of both columns are connected via a ribbon.

From both you can see a very clear message: Female employees were responsible for the majority of sales in December 2012, but in January 2013 this turned around in favor of the male employees. The 100% Stacked column chart is interactive, and you can ask for the exact numbers by hovering over with the mouse. Males were responsible for 1.3M of the sales amount in December 2012, which is 47.6 percent, and females for 1.4M (52.4 percent). This changed to 2.4M (57.9 percent) and 1.7M (42.1 percent) in January 2013. (No, I will not start a gender discussion here. This is just an example. And the data is rather random, taken from an AdventureWorksDW demo database. Even after working with this dataset for a decade, I didn't discover this fact until the Insights feature made me aware of it.)

If this result was only by chance, or if it was due to a certain strategy typical for our female and male employees, could be something to discuss with our peers. We can easily convert this insight into a "real" visual and add it to our report. Make sure to choose the 100% Stacked column chart and click on the plus (+) icon at the top-right corner of the insight. It will be added as a standard visual to the current report page, and we can change all properties. I changed the colors to traditional ones (females in red, males in blue), and I added a constant line (value 0.5) to show the 50 percent point, which makes it easier for the report user to see the change of majorities. You can see the enhanced report page in Figure 2-6.

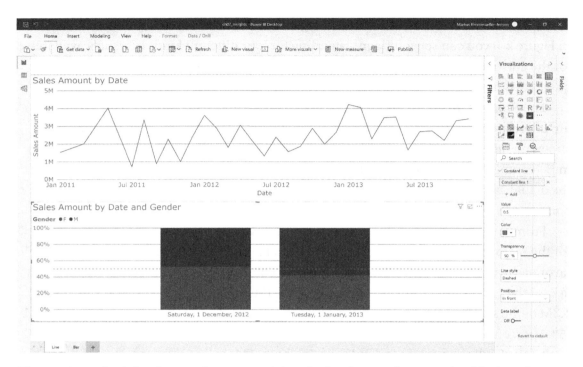

Figure 2-6. *Insight changed to a 100% Stacked column chart and added to the report with changed colors and an analytic line in green at 50 percent*

The Insights feature allows not only for finding out about an increase but also for finding an explanation for a decrease or different distributions. Find out more in the next sections.

Attention   At time of writing of this book, this feature was in preview. I do not recommend using preview features of any kind for analyses that are crucial for you or your organization. Preview features might be unstable, are unsupported, and might be changed fundamentally before they make it into the product, and you don't know when or if they make it into the product. Find updates on this feature here: `https://docs.microsoft.com/en-us/power-bi/desktop-insights`.

Explain the Decrease

If you right-click a data point in a line chart, where the value of the previous data point was higher, you get the choice of *Explain the decrease* via *Analyze*. I did this for the low data point at the end of the line chart in Figure 2-7 and received an *analysis of the 19.42% decrease in Sales Amount between Thursday, 1 August, 2013 and Sunday, 1 September 2013* and got the following list: Sales Amount by Data and ...

- English Promotion Name

- Product Line

- Gender

- Marital Status

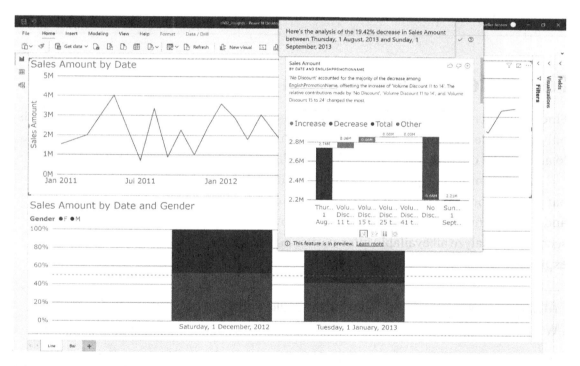

Figure 2-7. *Explain the decrease*

The first one (English Promotion Name) seems legit. It shows that *'No Discount'* *accounted for the majority of the decrease*, but that *Volume Discount 11 to 14* and *Volume Discount 15 to 24* softened the negative trend, as we sold more with these promotions than in the previous month. The rest of the features for *Explain the decrease* are identical to those for *Explain the increase*.

Attention   At time of writing of this book, this feature was in preview. I do not recommend using preview features of any kind for analyses that are crucial for you or your organization. Preview features might be unstable, are unsupported, and might be changed fundamentally before they make it into the product, and you don't know when or if they make it into the product. Find updates on this feature here: `https://docs.microsoft.com/en-us/power-bi/desktop-insights`.

Find Different Distributions

Finding an increase or decrease (over time) makes sense for continuous value on the axis only. If you use categorical data or a bar or column chart, the Insights feature will be able to analyze the distribution of values for the selected category over other categories.

In the example in Figure 2-8, I right-clicked the bar chart (it does not matter which bar you click or if you right-click on the background of the chart) and selected *Analyze* and *Find where this distribution is different*. Power BI Desktop will then take a few seconds to analyze all available information in your data and data model and suggest explanations that are statistically significant.

Attention   Not everything that is statistically significant is practically significant. Weigh the suggestions against your domain knowledge and your common sense.

Even when something has a practical significance it does not tell us directly which of the insights was the cause and which one was the effect. Again, we need domain knowledge and common sense to draw the right conclusions.

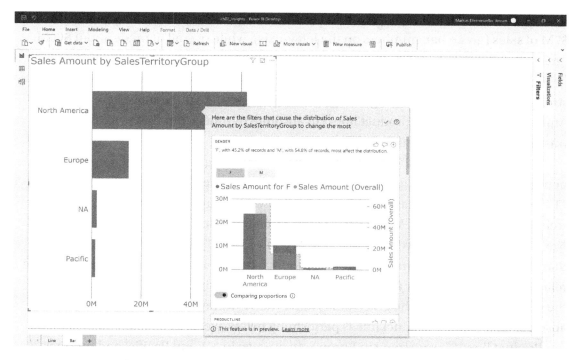

Figure 2-8. *Find where the distribution is different*

Power BI Desktop shows the insights in a separate dialog window with the following heading: *Here are the filters that cause the distribution of Sales Amount by SalesTerritoryGroup to change the most.* It came up with a list of categories:

- Gender

- Product Line

- Marital Status

- Date

- English Promotion Name

I find the second one, product line, the most interesting. The description tells us, *'T,' with 10.5% of records; 'S,' with 23.9% of records; and 'R,' with 33.3% of records, among others most affect the distribution.* Product line "T" is already pre-selected, and the green bars in the chart show us the distribution of this product line over the different regions (with the values on the left y-axis), filtered for product line "T." The grey line chart shows the overall sales amount (independent of any product line or any other category)—with the values shown on the right y-axis. In "North America," this product line made about $7M of sales (green bar, numbers on the left y-axis), while overall, it made about $60M (grey line, numbers on the right y-axis).

The chart in Figure 2-8 leads us to the conclusion that sales for North America are strong for all product lines (height of green bar and grey line is the biggest for North America in the case of all product lines) in absolute numbers, but that product line "T" was relatively weak in North America.

Why is product line "T" weak, you might ask? The insight lies in the comparison between the height of the green bar and the grey line per region. If sales for product line "T" would have been average, then the green bar and the grey line would be of the same (relative) size in the chart. When you see the green bar being shorter than the grey for North America, that tells you that sales for product line "T" have been below average in that region. The relative share of sales for product line "T" in North America is below the relative share of sales over all product lines in North America. The opposite is true for the other regions, where the green bar is taller than the grey line. So, despite the strong absolute number, the product line is performing below average.

Product line "S" was strong in Europe, on average in NA and Pacific; and product line "R" was comparably strong in North America, on average in NA, and weak in Europe and Pacific. Only because of product line "R" is North America is so strong as a market. Without product line "R," the difference in absolute numbers between North America and the other regions would not be as big as it is currently.

This brings us to the question of why this product line is so strong in the non–North American market. Or why it is so weak in the North American market itself. If you want to discuss this insight with your peers, you can simply add the current chart to your report page by clicking on the plus (+) button, as I have done in Figure 2-9.

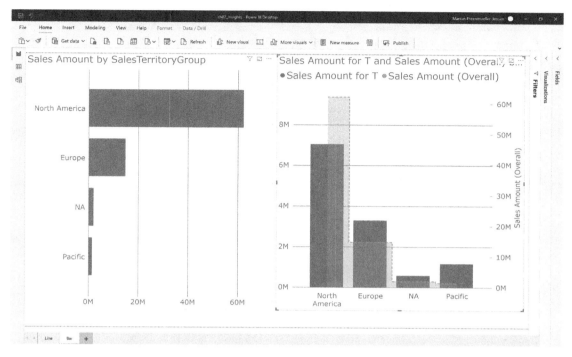

Figure 2-9. *Distribution for product line "T"*

When you click the yellow *Comparing proportions* button (Figure 2-8), the button color changes to grey and the text to *Comparing absolute values,* as you can see in Figure 2-10. And that's exactly what we get from the chart then: We now have only one single axis from which we read the values for both the product line's sales amount and the overall sales amount. The proportions are now harder to see, but it's easier to see that the $7M of sales in North America for product line "T" is small compared to the overall number of $60M.

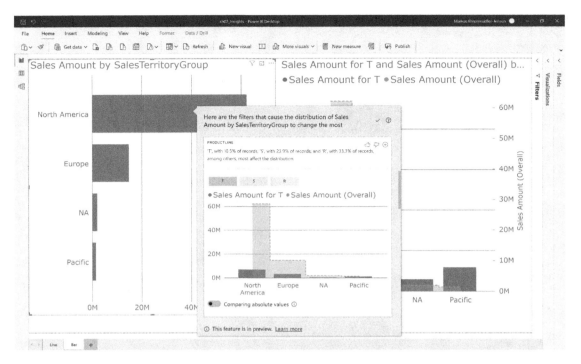

Figure 2-10. *Distribution based on absolute values*

Types of Insights

The features just described are not available for custom visualizations. Only the following standard visualizations support the Insights feature in Power BI Desktop. Right-clicking and selecting *Analyze* will show both or one of these:

- *Explain the increase/decrease*: Line chart, area chart, stacked area chart; only when you have a continuous category, like date/time, on the axis

- *Find where this distributions is different*: Stacked bar chart, stacked column chart, clustered bar chart, clustered column chart

If you put a continuous category into a bar or column chart, you can *a*nalyze both the increase/decrease and different distributions. But I don't see much sense in creating such a visual in the first place, as best practice is to visualize continuous data on a line chart.

Note Examples for continuous data include date, time, percentage, quantity, amount, and so forth. Theoretically, there is an endless number of valid values between every two numbers you are coming up with (between 1 and 2 there can be, for example, 1.5).Examples for categorical data include non-numerical data, like a product category or a sales territory, but can include numerical data as well. Think of a rank, which is numeric, but there is no space between the first and the second position in a race, as there is no position 1.5 possible.

Quick Insights Feature

Power BI Desktop is only one of the tools available in the Power BI suite. We can publish reports in the Power BI Service, which is hosted by Microsoft's Azure cloud data centers. You can consume published reports in an internet browser (without having Power BI Desktop installed on your device). This is possible even if we are using Power BI free of cost (without having payed for a Pro license or a Premium capacity in the cloud; you need the latter two if you want to share the report in the service with other people or use enterprise features). Here is a description how to find out which type of license you own: `https://docs.microsoft.com/en-us/power-bi/consumer/end-user-license`.

In the *Home* menu you can click *Publish*, which is the only button inside the *Share* section at the very left end, to upload the report to the Power BI Service (Figure 2-11).

Figure 2-11. *Select Home ➤ Publish to upload the file to Power BI Service*

You don't need to log in with an account to use Power BI Desktop features. But you need a login if you want to publish your report to the Power BI Service. Unfortunately, a) free accounts (like outlook.com or hotmail.com) will not work and b) the administrator of the domain has to enable the possibility to use it as an account for Power BI. Please refer to `https://docs.microsoft.com/en-us/power-bi/fundamentals/service-self-service-signup-for-power-bi` for further details.

If you don't have a Pro license, the only workspace you can upload to is "My workspace," as is the case in Figure 2-12.

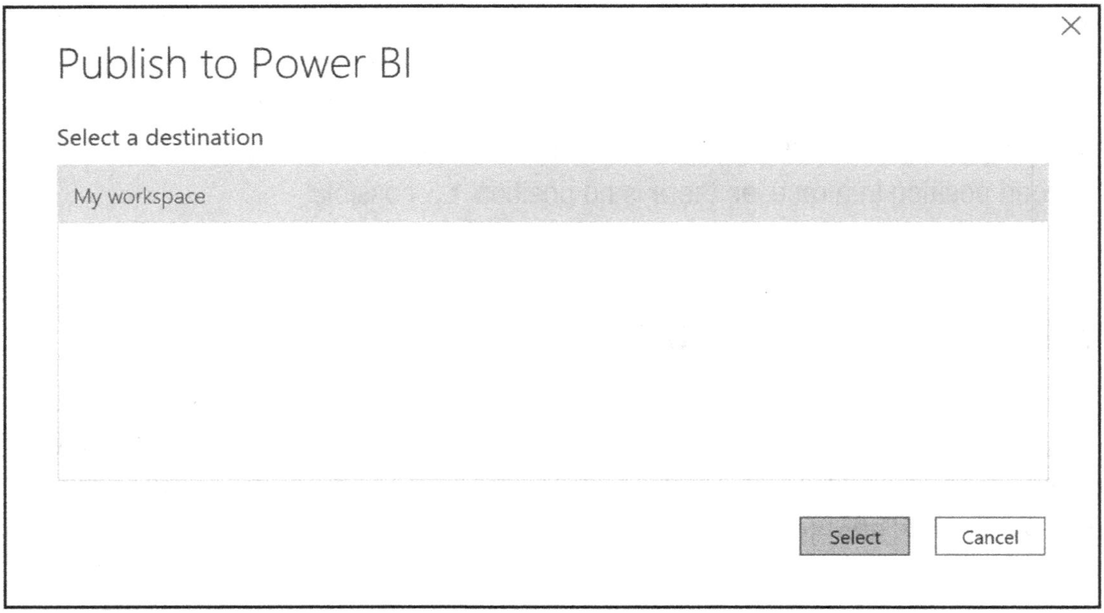

Figure 2-12. *Select a destination workspace when publishing to Power BI Service*

If a report with the same name already exists in the selected workspace, a dialog box (Figure 2-13) will open, asking you if you want to replace it. If you do not want to replace it, you can press *Cancel* and either rename the report in Power BI Service or save the current file under a different name in Power BI Desktop before you publish the report again.

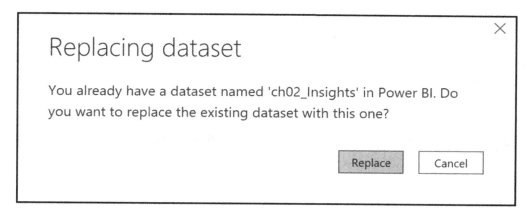

Figure 2-13. *Either replace an existing dataset or cancel and save the current file under a new name first*

Depending on the size of the file, the bandwidth of your internet connection, and how busy Power BI Service is at the moment, the upload may take a few seconds or longer.

As soon as the report is successfully uploaded, you can open it directly from the dialog box seen in Figure 2-14 by clicking on *Open 'ch02_Insights.pbix' in Power BI*. The other option is to *Get Quick Insights*, but this does not always work for me (as it opens the list of dashboards instead of Quick Insights).

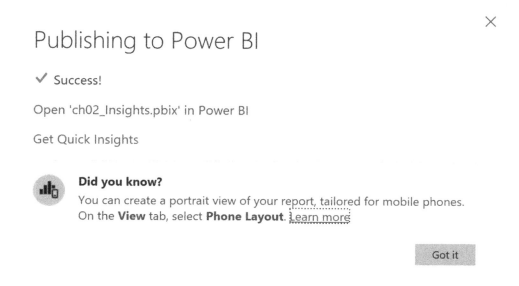

Figure 2-14. *After the file is published, you can open the report in Power BI Service or directly get Quick Insights*

Either way will lead you to the Power BI Service at `app.powerbi.com`. And there you can select *My workspace* on the left side of the screen and then select *Reports* in the middle of the screen. Click on the lightbulb icon in the *Actions* section to the right of the name of your report, as marked in Figure 2-15.

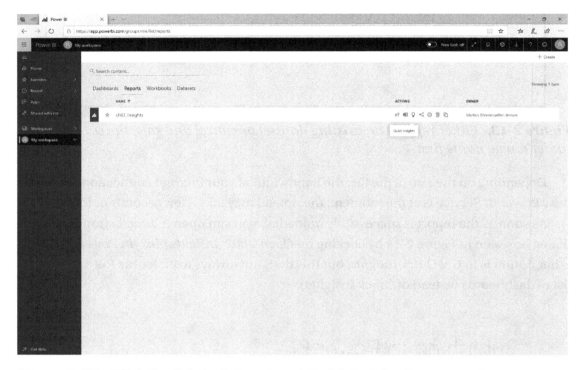

Figure 2-15.   *Click the lightbulb icon to get Quick Insights from your data*

If you haven't activated this feature since you last uploaded the file, Power BI Service will start searching for insights for you. It will analyze the whole content of the uploaded Power BI file to find statistically significant insights (Figure 2-16).

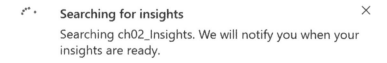

Figure 2-16.   *Power BI Service applies machine learning algorithms against the data model to find statistically relevant insights*

As soon as the insights are found, you can click *View insights,* as in Figure 2-17. The next time you click on the lightbulb icon, shown in Figure 2-15, you will go directly to the Quick Insights (Figure 2-18).

Figure 2-17. *View insights directly after they are ready*

Typically, Quick Insights in the Power BI Service comes up with a much longer list of insights than does Power BI Desktop—and they are more specific, with no options to change the visualization in the overview. You can see the first ones in Figure 2-18.

Figure 2-18. *Quick Insights for the demo file*

The feature usually shows forty insights, from which I want to highlight the following Quick Insights:

- *Sales Amount by English Promotion Name* as ring chart (and later again as a bar chart) with the Quick Insight "No Discount accounts for the majority of Sales Amount." We usually sell our goods without any promotional discount.

- *Average of Unit Price Discount Pct by Product Line* as bar chart with Quick Insight "'T' has noticeably more Unit Price Discount Pct." Product line "T" is the one where we give an almost 2 percent discount (on average) compared to less than 1 percent on the other product lines.

- *Average of Standard Cost by Date* as line chart with Quick Insight "Standard Cost for Product Line 'S' has outliers." Costs for this product line are volatile and can be below $9 and above $23.

- *Count of Employee and Count of Territory by English Promotion Name* as correlation chart with the Quick Insight "There is a correlation between Employee and Territory." The more territories a region has, the more employees we have there.

- *Average of Unit Price Discount Pct and Average of Unit Price by Date* as scatter chart with Quick Insight "Unit Price Discount Pct and Unit Price form cluster when grouped by Date except for Tuesday, 1 November, 2011." On the one hand, we can recognize a yellow cluster, which has no discounts and is on the upper end of the price scale. On the other hand, we have a grey cluster with lower prices, but discounts on some days. When you move the mouse over the red data point, it reveals that on this day we gave a 6 percent (0.06) discount, which is indeed unusual. This day appears in other Quick Insights for similar reasons.

- *Sales Amount by Product Line* as bar chart with the Quick Insight "'R' and 'M' have noticeably more Sales Amount." Product Lines "R" and "M" are our best sellers.

- *Count of Sales Territory by Date* as line chart with Quick Insight "Sales Territory for Product Line 'S' shows several trends." This product line sold in just a few territories in the beginnings of 2013 and 2014, but in more areas in the time between.

- *Average of Standard Cost by Date* as line chart with Quick Insight "Standard Cost for Product Line 'R' has outliers." The standard costs for this product line area volatile, but are lowered each year.

- *Sales Amount and Average of Unit Price by Date* as scatter chart with Quick Insight "Sales Amount and Unit Price form clusters when grouped by Date." The yellow cluster consists only of data points in year 2011. We succeeded in increasing sales amount by increasing prices (or selling more expensive goods) from 2012 onward.

- *Count of Promotion and Sales Amount* as scatter chart with Quick Insight "Promotion and Sales Amount have outliers for Date Sunday, 1 May, 2011, Friday, 1 July, 2011, and Monday, 1 August, 2011." July is special, as we did give only one single promotion. May and August are special, as we gave discounts on very high sales amounts.

Even given the fact that the dataset used is an artificial one, with no real-life connection, these insights give us a quick idea of what kind of interesting relationships are hidden in the dataset. Usually at least some of the Quick Insights are a good starting point for further analysis.

With each of the Quick Insights you have got two options:

- Click the chart anywhere (except for the pin icon) and enter focus mode (Figure 2-19). This will expand the chart onto the whole available space, making it easier to distinguish smaller differences. (Click *Exit Focus mode* on the top left to return to the previous overview of all Quick Insights.)

- Click the pin icon to pin this Quick Insight to a Power BI Service dashboard. A dialog will then ask you if you want to create a new dashboard or pin the chart to an existing one. Power BI dashboards are out of scope for this book. Find more information about dashboards here: `https://docs.microsoft.com/en-us/power-bi/service-dashboards`.

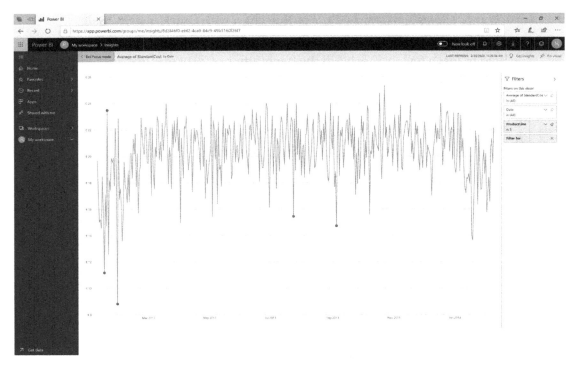

Figure 2-19. *Quick Insight Average of Standard Cost by Date in focus mode*

Types of Quick Insights

Power BI Service is capable of finding Quick Insights for you based on correlation and time-series analysis. Here is a list of things Power BI Service can discover for you in the data model of your Power BI Desktop file:

- Majority (e.g., most of the sales amount is via a certain category)

- Outliers (categories that have bigger or smaller values compared to the average)

- Steady share (when a share of subcategories within a category stays the same)

- Low variance (e.g., the unit price for all products in a certain category is almost the same)

- Trends (upward or downward), seasonality (periodic patterns over weeks, months, or years), trend outliers (when the upward or downward trend is interrupted), and change points over time (when a trend turns around once or more often)

- Correlation (this can be positive, e.g., the higher the costs, the higher our list price; or negative, e.g., the higher the costs, the lower our margin).

Key Takeaways

In this chapter, you learned the following:

- A continuous value on the axis allows you to ask Power BI Desktop to *e*xplain the increase or to *e*xplain the decrease (when you right-click a data point and select *Analyze*).

- In bar and column charts you can *a*nalyze via a right-click and let Power BI Desktop *f*ind where distribution is different.

- You can add a discovered insight to your report page.

- Power BI Service offers a similar feature called Quick Insights, which is capable of finding more complex insights in the whole dataset (independent of any visual), which you can add to a dashboard in the Power BI Service.

CHAPTER 3

Discovering Key Influencers

The Insights and Quick Insights features (see Chapter 2) discover statistically significant insights on their own. The Key Influencers visual gives you more control over that approach, as it allows you to explicitly specify the columns involved. This visual will help you to discover characteristics of a field of your choice. Behind-the-scenes machine learning algorithms are automatically looking for correlations in your data to show you what attributes are making, for example, your Bike product category so different from Accessories. You bring the field value, categories, and measures; Power BI will bring the insight into how those categories and measures are key influencers on the field value.

Introduction

In Figure 3-1 you see the Key Influencers visual already in action. First, we will concentrate on the Key Influencers part of this visual (underlined in green); later, we will work with the Top Segments part.

© Markus Ehrenmueller-Jensen 2020
M. Ehrenmueller-Jensen, *Self-Service AI with Power BI Desktop*, https://doi.org/10.1007/978-1-4842-6231-3_3

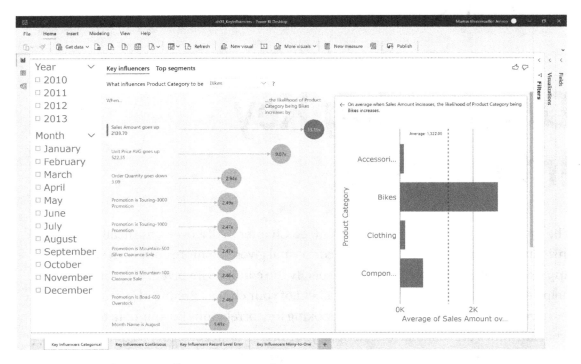

Figure 3-1. *The Key Influencers visual*

The visual does a great job summarizing what it is showing to us in plain English. Right on the top it states the question, which will be answered in the two visuals below: *What influences Product Category to be Bikes?*

The answer is split into two sections below. On the left, we see different measures and categorical data that have a statistically significant influence on product category Bikes. The list contains the following:

- Sales Amount

- Unit Price AVG

- Order Quantity

- Promotion

- Month Name

Attention   Not everything that is statistically significant is practically significant. Weigh the suggestions against your domain knowledge and your common sense.

Even when something has a practical significance it does not tell us directly which of the insights was the cause and which was the effect. Again, we need domain knowledge and common sense to draw the right conclusions.

Just above this list we have *"When ..."* and *"... the likelihood of Product Category being Bikes increases by."* Together with the preceding list entries we get sentences like *"When Sales Amount goes up 2139.70 the likelihood of Product Category being Bike increases by 13.19x."*

On the right there is a bar chart that gives us more details about the 13.19x increased likelihood. Again, we have a description in plain English at the top: *"On average when Sales Amount increases, the likelihood of Product Category being Bikes increases."* The bar chart shows the *Average of Sales Amount over Reseller Sales.* Hovering the mouse over the bar at *Bikes* shows us that for this category the *Average of Sales Amount* is 2,673.48. We sold products from category Bikes exactly 24,800 times. The red line shows that the *Average of Sales Amount* over all categories is only 1,322.0. Obviously, bikes are more expensive than the other product categories and therefore are giving us a higher *sales amount* on average per sale.

The visualization on the right section depends on what you have selected in the left part. In Figure 3-1, *Sales Amount* is selected. *Sales Amount* is marked with a green bar on its left. And the bubble for 13.9x is green as well, as opposed to the grey bubbles for the other list elements. When you click on the second element, you get a similar chart on the right, for the *Unit Price AVG* instead of *Sales Amount.* The type of visualization on the right changes though, when you select *Promotion is Touring-3000 Promotion* (which increases the likelihood 2.49x). The right visualization changes to a column chart. The column for said promotion is colored in green, the others in blue.

Analyze Categorical Data

When you ask the Key Influencers visual to analyze categorical data (as we have done), it will give you a listbox to further clarify which category value you want to be analyzed. I added *Product Category* to *Analyze*, as you can see in Figure 3-2. We then must decide which of the available values of *Product Category* should be analyzed: *Accessories, Bikes,*

Clothing, or *Components.* This is done in the top line of the visual (*"What influences Product Category to be ..."*). Each category is influenced differently by the fields listed in the *Explain by* section. As soon as you change your choice, the Key Influencers visual will re-run its analysis to find statistically significant influencers.

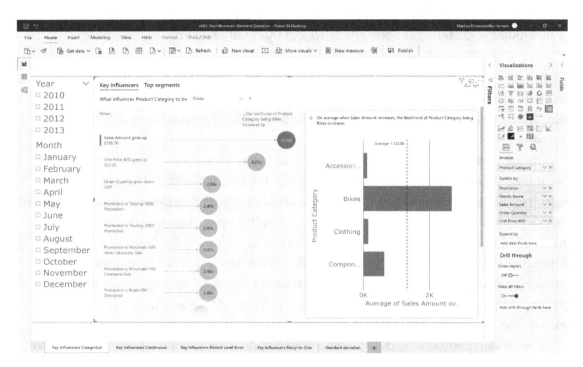

Figure 3-2. *The Key Influencers visual is asked to analyze the field Product Category*

From the example we can learn that *a*ccessories are sold mostly in June (1.27x) and January (1.14x), while *b*ikes are sold mostly in August (1.41x), July (1.25x), May (1.12x), June (1.12x), and October (1.08x).

Analyze Continuous Data

Continuous data does, by definition, have an endless amount of possible values. It therefore does not really make sense to select a certain value for which the Key Influencers visual would then look for influencers (as for most of the possible values of the continuous field no or not enough examples would be present to make the analysis work). Instead of a list of available values we only can select between *Increase* or *Decrease*.

I created a new report where I exchanged columns *Sales Amount* and *Product Category* with each other: I put *Sales Amount* into the *Analyze* section and *Product Category* into the *Explain by* section, as you can see in report *Key Influencers Continuous* in Figure 3-3. The result of this analysis is then a list of what has a positive or negative correlation with the field in *Analyze*. The rest of the capabilities are the same as for categorical data.

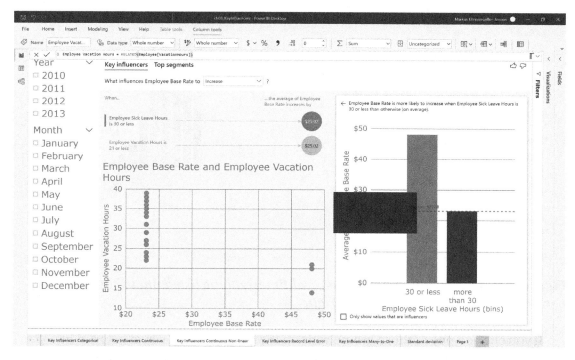

Figure 3-3. *For continuous data you can analyze influencers that make the variable increase or decrease*

Explain by Categorical Data

Add fields to the *Explain by* section that you think could have an impact on the field sitting in *Analyze*. These can be of either categorical or continuous data type. You see examples of different fields above in Figure 3-2.

For categorical data in the *Explain by* field, the Key Influencers visual evaluates each category and shows the most influential ones. In Figure 3-4 you can see that *Promotion* is listed not only once, but several times. Bikes were often sold through promotions *Touring 3000 Promotion, Touring 1000 Promotion, Mountain-500 Silver Clearance Sale, Mountain-100 Clearance Sale,* or *Road-650 Overstock.*

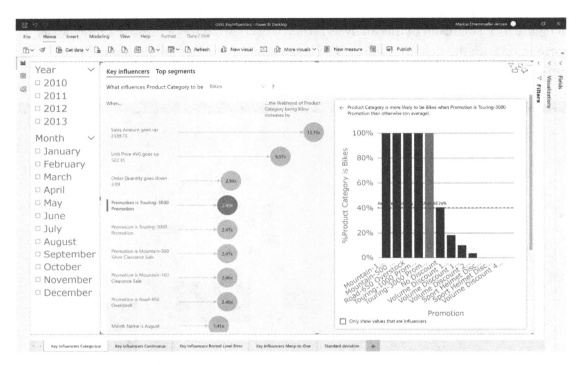

Figure 3-4. *Promotions are listed several times. All the listed ones show 100 percent in the column chart on the right*

If you click on one of those promotions on the left part of the visual (as I did in Figure 3-4), you will see that the columns for those promotions are 100 percent. That means that those promotions were solely applied onto sales for bikes. This explains the high influence.

The column chart on the right of the visual is the same for all listed promotions, except the current column, marked in green, changes. Try this out on your own.

If you click on checkbox *Only show values that are influencers,* all promotions not listed on the left will be removed from the column chart (Figure 3-5).

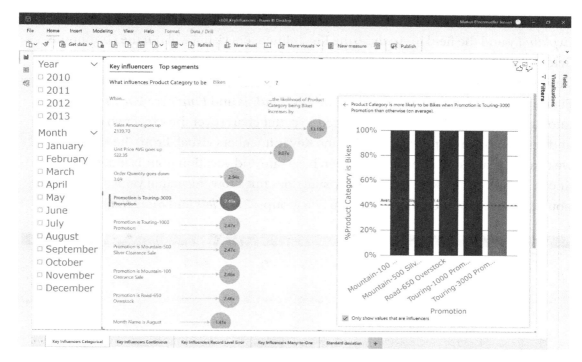

Figure 3-5. *"Only show values that are influencers" shows only influencers in the right chart*

Categorical data in *Explain by* leads to a column chart, with the values of this field on the x-axis. The y-axis depends on the data type of the field in *Analyze*. If this is of categorical data type as well, then you get columns, with the height representing the percentage of how many rows explain the selected value for the field in *Analyze*. The average over all categories is drawn as a red line.

If the field in *Analyze* is of continuous data type, like column *Sales Amount* above in Figure 3-3, then the y-axis represents the average of this field, per value of the field in *Explain by*, which are shown on the x-axis. The average over all categories is drawn as a red line.

Explain by Continuous Data

If you put continuous data into the *Explain by* field, the Key Influencers visual calculates the standard deviation for this data field and shows this in the left visual. In Figure 3-2 we are told that if *Sales Amount* goes up by 2139.70 the likelihood of *Product Category's* being Bikes increases by 13.19x. The amount of 2139.70 is the standard deviation.

In cases where there is non-linear correlation between the continuous field in *Explain by* and the field in *Analyze*, the Key Influencers visual will *bin* the continuous field. That means it will group the values of the field in *Explain by* into buckets and treat those buckets like categorical data (see "Explain by Categorical Data" section). In Figure 3-6 I've added *Employee Base Rate* into *Analyze* and *Employee Vacation Hours* into *Explain by*. Additionally, I've created a scatter chart (over the exact two columns) and put it into the empty space inside the Key Influencers visual. From the scatter visual we can see that the relationship between base rate and vacation hour is indeed non-linear. Therefore, the Key Influencers visual does not show the actual values of vacation hours but rather binned the values into two groups: *21 or less* and *more than 21*.

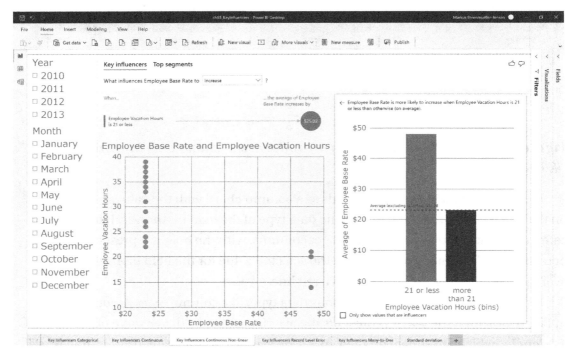

Figure 3-6. *Employe Base Rate and Employee Vacation Hours have a non-linear correlation, which is shown in an extra scatter plot positioned above the Key Influencers visual*

Continuous data in *Explain by* leads to either a bar chart or a scatter chart. You get a bar chart, with the average values of the field in *Explain by* on the y-axis, when the field in *Analyze* is of a categorical data type (or of a continuous data type but binned because of the reasons just described). The possible values for the field in *Analyze* are shown on the y-axis. The average over all categories is drawn as a red line.

When you combine fields of continuous data type in both *Analyze* and *Explain by* and there is a linear correlation, then you end up with a scatter plot on the right. The possible values of the field in *Explain by* are shown on the x-axis. Either the actual or the average value of the field in *Analyze* is shown on the y-axis, depending on the chosen *Analysis* type in the format options (see further below). This time, the red line represents a simple regression line over the shown data points.

Setting Granularity

The Key Influencer visual is looking for correlations in your data. The correlation of variables is dependent on the granularity of your analysis. A correlation might not exist on the row level of your transactional data (e.g., per row in table *Reseller Sales*), but a strong correlation might exist when the values are aggregated per product category. Or it might be the other way around. The granularity level the Key Influencer visual is working on is defined by what fields you are using in *Analyze*, *Explain by*, and *Expand by*. We have already discussed *Analyze* and *Explain by*, but I did not mention that they influence the granularity as well. The sole purpose of *Expand by* is to set the granularity level without having the Key Influencer visual looking for correlations of these fields to the field in *Analyze*.

In Figure 3-7 you see two table visuals and two scatter plots. All four have the same content: *Sales Amount*, *Product Category*, *Order Quantity*, and *Unit Price AVG*. But the visuals show totally different numbers: The two upper ones (titled *Sum Sales Amount*) only show four entries, one per product category. The lower ones (titled *Don't summarize Sales Amount*) show lower numbers in each entry, but many more entries.

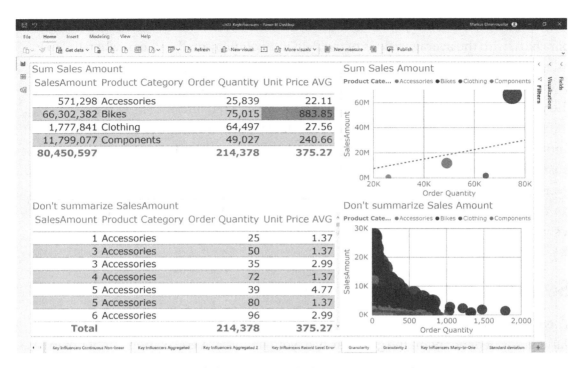

Figure 3-7. *Different granularity gives a different picture of the data*

The explanation is that all show the same data, but on different levels of granularity. The difference lies in the summarization property of column *Sales Amount*.

In the upper ones it is set to *Sum*. Therefore, *Sales Amount* is aggregated (summed) per product category. In the lower ones it is set to *Don't summarize*. Therefore, *Sales Amount* is not aggregated, but every single row of the table the column belongs to (*Reseller Sales*) is shown. It's the same data, but with different pictures and different conclusions per level of granularity.

So, we learned that we can influence the level of granularity by setting the summarization settings of a numeric column. What if we want to use Key Influencer visual on a measure instead? Glad you asked! That's where the *Expand by* field list comes into play. A measure is always aggregated (and performs like a numeric column with summarization turned on). To change the granularity to, for example, the row level of *Reseller Sales*, I put *SalesOrderNumber* and *SalesOrderLineNumber* into the *Expand by* field list. The combination of both columns uniquely identifies a row in table *Reseller Sales*. (We could therefore call the combination an index, a key, a row identifier, or an ID.) In Figure 3-8 you see a report containing four Key Influencer visuals. The two on top and the two on bottom show the same information each. The two on the left are built with numeric column *Sales Amount*, once with summarization set to *Sum*, once with summarization set to *Don't summarize*. The two on the right are built with measure *Sales Amount*. For a measure we cannot change the summarization—the visual on top right (with the measure) behaves identically to the one on top left (numeric column with summarization set to *Sum*). To achieve the same behavior for a measure as with a numeric column that is set to *Don't summarize* (see bottom left in Figure 3-8), I added *SalesOrderNumber* and *SalesOrderLineNumber* into *Expand by*.

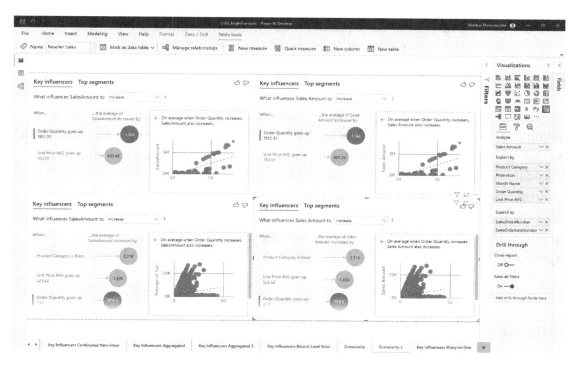

Figure 3-8. *Expand by demonstrated*

Which one is the right granularity level for the Key Influencer visual? I would tend to say, the level of granularity that you have control over—which will be mostly the lowest granularity available.

> **Note**   If a table does not contain a row identifier (like the combination of
> *SalesOrderNumber* and *SalesOrderLineNumber*) you can create one explicitly in
> Power Query. Select *Home* ➤ *Transform data* in the ribbon of Power BI to open
> Power Query. Select the query and then *Add Column* ➤ *Add Index Column* from
> the ribbon in Power Query.Please be careful though, as an index column cannot be
> compressed very well. The size of your PBIX file may increase significantly, and the
> performance may suffer.

Filters

The Key Influencers visual, like all other visuals, is responding to filters. This enables us
to find out if the influencers are different for different portions of the data available. With
no filter applied, we found out that bikes are mostly sold in the summer months: May,
June, July, August, and October (Figure 3-1).

This is even more true for year 2013. If you select this year on the slicer next to the
Key Influencers visual, you will find out that in this year only June, July, and August were
strong months, and therefore show up as influencers in the visual, which you can see in
Figure 3-9.

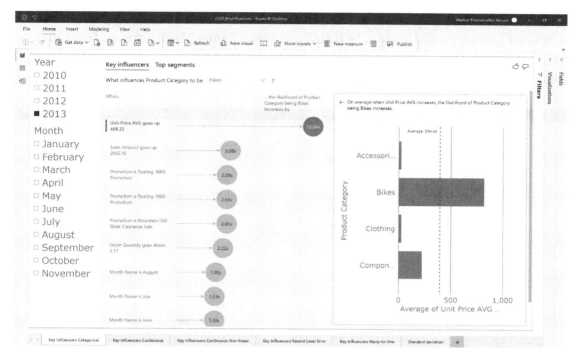

Figure 3-9. *Filters limit the data available for the Key Influencers visual and allow you to analyze different portions of the available data*

Note   Power BI Desktop offers a bunch of different possibilities to filter a visual: direct filters on the selected visual (in filter pane), filters that apply on all visuals for the current page (in filter pane), filters that apply on all visuals for all pages in the current file (in filter pane), and cross filters that interact with other visuals on the same page (including from a slicer visual, like in Figure 3-9).

Top Segments

The *Key Influencers* feature of the Key Influencers visual (see above) looks at the influence of one individual field per time. The *Top Segments* feature enhances this analysis to look at combinations of fields (you have put into *Explain by*).

When I selected *Top Segments*, like in Figure 3-10, and leave the question as, "*When is Product Category more likely to be Bikes?*" the visual "*found 4 segments and ranked them by % Product Category is Bikes and population size.*" Before we go on to see more details per segment, let's look into those four segments.

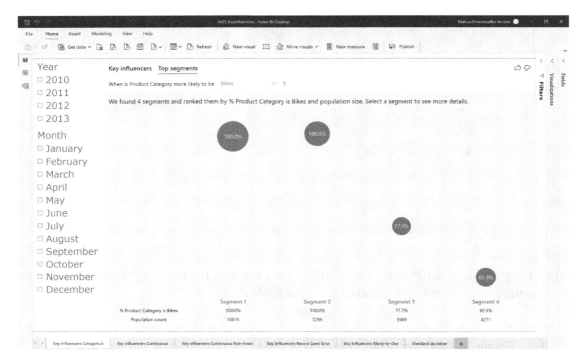

Figure 3-10. *Besides key influencers, this visual can also discover top segments*

The segments are generically named and numbered (*Segment 1* to *Segment 4*) and are ordered by importance: *Segment 1* consists purely of bikes *(% Product Category is Bike* is *100%*) and a *Population count* of *10615*, which you can read from the bottom of the visual. That means that the Key Influencers visual could find a segment containing 10,615 rows inside the *Reseller Sales* table (that's the table containing the *Product Category* field) where bikes were sold. *Segment 4*, on the other hand, consists of 65.9 percent bikes (the rest are other product categories) and contains 4,271 rows.

The two numbers (*% and Population count*) influence the size and height of the green bubbles: The higher the percentage, the higher the bubble. The bigger the population count, the bigger the bubble (which is easy to remember). That's why the bubble representing *Segment 3* is positioned higher but is drawn (slightly) smaller than the bubble for *Segment 4*.

In Figure 3-11, I put *Sales Amount* into *Analyze*. This is a field of continuous data type, and the top segment picture looks a little bit different. We can choose between *"When is SalesAmount more likely to be High?"* and *"When is SalesAmount more likely to be Low?"* In the chart below the visual *"found 4 segments and ranked them by Average of SalesAmount and population size,"* we still get the *Population count*, but instead of percentages of likelihood we get concrete numbers. *Segment 1* explains an *Average of SalesAmount* of 8.06K through a total of 3,228 rows. *Segment 4* only an *Average of SalesAmount* of 1.52K, but for 5,737 rows.

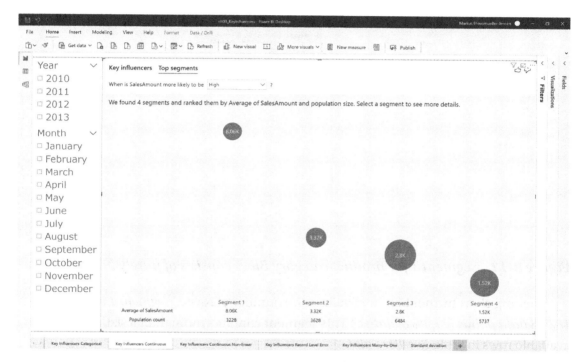

Figure 3-11. *When analyzing continuous fields, you can select either High or Low (instead of a concrete field value)*

Top Segments Detail

In Figure 3-10 I clicked on Segment 1 (either click on the bubble or on text on the bottom of the visual). Figure 3-12 reveals that this segment consists of only a single influencer: rows with a *Unit Price AVG* of greater than 858.9. On the right we get a description in plain English: *"In segment 1, 100.0% of Product Category is Bikes. This is 59 percentage points higher than average (40.8)."* That means that across all our sales, 40.8 percent were *bikes*, but *Segment 1* consists of purely (= 100 percent) bikes. This is visualized as two bar charts just below. If a sold product has a *Unit Price AVG* of greater than 858.9, it is a bike for sure.

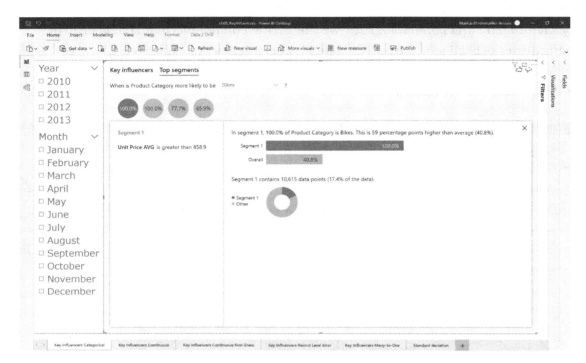

Figure 3-12. *Segment 1 for product category Bikes consists of purely bikes*

The ring chart further below visualizes the number of rows. "*Segment 1 contains 10,615 data points (17.4% of data).*" This segment contains more than a sixth of all available rows in table *Reseller Sales*.

In Figure 3-13 I ask for "*When is SalesAmount more likely to be High?*" and get a *Segment 1* that shows in the details "*Order Quantity is greater than 3*" and "*Unit Price AVG is greater than 874.794.*"

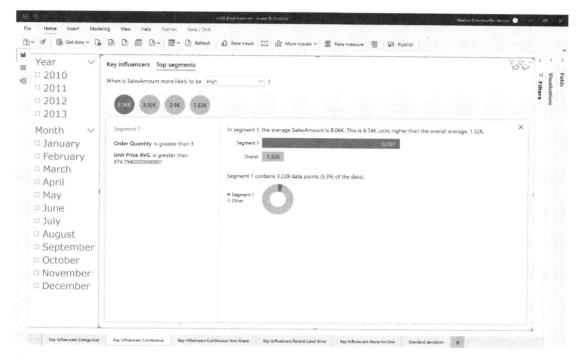

Figure 3-13. *Segment 1 for high values of Sales Amount*

When I turn the question around and ask for *"When is SalesAmount more likely to be Low?"* it comes up with a *Segment 1* that consists of *"Order Quantity is less than or equal to 3"* and *"Unit Price AVG is less than or equal to 37.25"* (Figure 3-14).

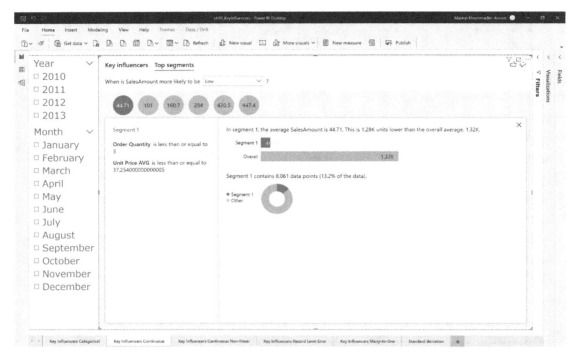

Figure 3-14. *Segment 1 for low values of Sales Amount*

Those segments on *Sales Amount* might not be of such great insight: Basically, it means that we have high sales when we sell an expensive product (high unit price) many times (high order quantity). And that we have low sales when we sell a cheap product only up to three times.

I'm sure you can come up with more useful segments by playing around with the fields in *Explain by*, with either the sample file or your own dataset.

Types of Influences

Not all added fields are shown in the visual. From the added ones, only those that actually have a statistical significance on the field we put into *Analyze* are shown. The number of rows per category in *Explain by* is considered as well. This means that a category with only a few rows might not show up as an influencer.

Fields

This visual offers the following fields (Figure 3-2):

- *Analyze*: You can only put one single field here. The visual looks for influencers on this field.

- *Explain by*: You can put as many fields as you want here. Amongst those fields, the influencers on the field in *Analyze* are looked for.

- *Expand by*: You can put as many fields as you want here. These fields define the granularity of an (implicit) measure you have put into *Analyze*. Usually you would put a key field in here. This field is not evaluated as a potential influencer.

Format Options

In addition to the "usual" options you have for most of the visuals (Title, Background, Lock aspect, General, Border, and Visual header), you can tune the behavior of the Key Influencers visual through the following categories:

- Analysis

- Analysis visual colors

- Drill visual colors

Analysis (Figure 3-15):

- Enable key influencers: Turns selection of *Key influencers* in the visual on and off.

- Enable segments: Turns selection of *Top segments* in the visual on and off.

- Analysis type: Can be *Categorical* or *Continuous*. Find more information later.

- Enable count: Adds a grey ring around the percentage bubble in the Key Influencers view representing the percentage of rows that the influencer contains. Examples are listed below.

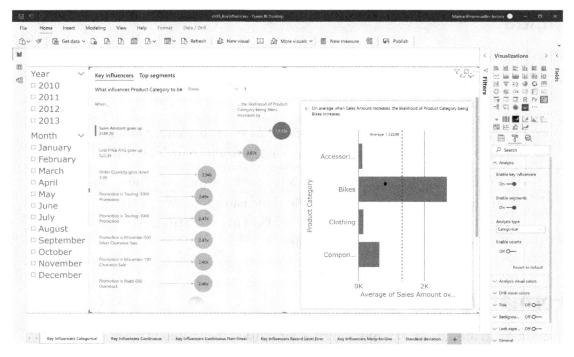

Figure 3-15. *Analyis section of the format options*

Setting the *Analysis type* to *Categorical* drives the Key Influencers visual to find reasons for a specific value for the field in *Analyze*. If you set this to *Continuous* (which is only available for fields of numeric data type in *Analyze* like *Sales Amount* but not for *Product Category*), the Key Influencers visual will try to find reasons why value for the field in *Analyze* increases or decreases. Figure 3-16 shows report *Key Influencers Continuous*, where I put *Sales Amount* into *Analyze* and *Analysis type* is set to *Continuous*. Therefore, I can only select *Increase* or *Decrease* in the listbox in the top region of the visual.

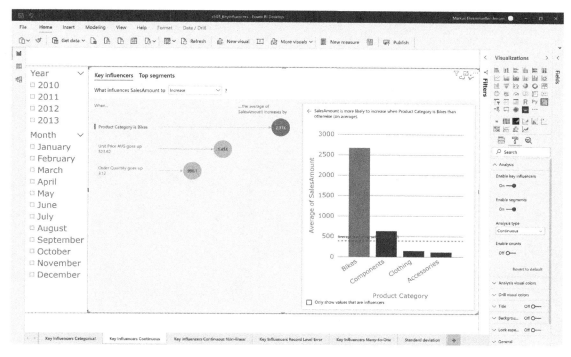

Figure 3-16. *Analyis type is set to Continuous*

In Figure 3-17 I changed this to *Categorical* in report *Key Influencers Continuous*. This has two effects: One, I get the warning "*SalesAmount has more than 10 unique values. This may impact the quality of the analysis.*" And second, I can now choose a specific value for sales amount. As there is not enough data for a single sales amount (that means we did not have enough sales for the sales amount selected), statistical analysis does not make much sense, and in most cases we do not see any influencers, but a warning instead: "*SalesAmount is 1 does not have enough data to run the analysis.*" If you try often enough, you will find individual sales amounts with enough rows, though (e.g., 1,017 worked for me).

Figure 3-17. *Analyis type is set to Categorical*

When *Enable count* is turned on, a grey ring is added to the green bubbles in the *Key influencers* overview. The calculation and length of the ring depends on the *Count type* property, which is only visible if you enable the count. You can choose between *Absolute* and *Relative.* Set the *Count type* to *Absolute* if you want the ring to represent the absolute number of rows. A full ring would than represent 100 percent of the data available for the visual. None of the key influencers in Figure 3-18 do show a full ring—none cover all rows.

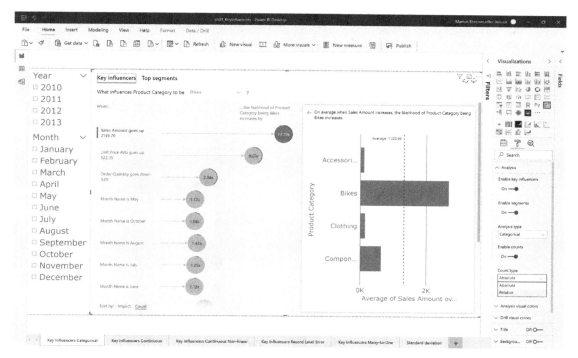

Figure 3-18. *Count type is set to Absolute*

Choose *Relative* if you want a full ring for the biggest influencer so you can compare how many rows are represented by the other influencers compared to the biggest one. In Figure 3-19, the first three influencers (*Sales Amount, Unit Price AVG*, and *Order Quantity*) have a full ring. From the number of rows, those influencers cover the most rows. The next categories (months) only cover about a fourth of the rows of the first three. Therefore, the ring only covers a fourth of the bubble.

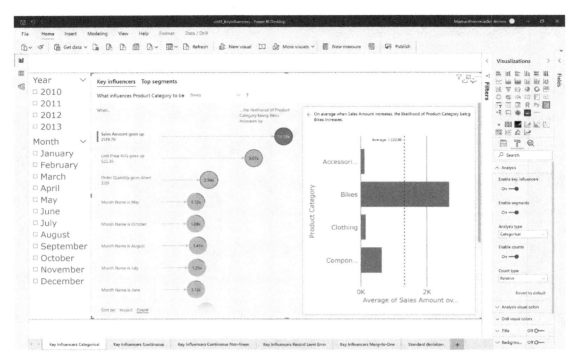

Figure 3-19. *Count type is set to Relative*

When *Enable count* is turned on, the bubbles are *Sorted by Count* of rows (that means the size of the grey ring). You can change this back to *Impact* (that means the number inside the bubble) if you prefer the original order (Figure 3-20).

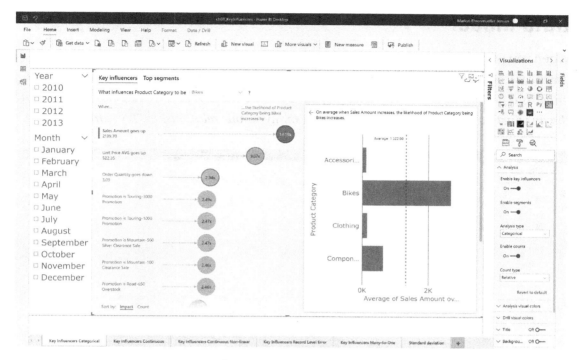

Figure 3-20. *Show the counts (as rings), but order the influencer by impact*

Analysis visual color lets you set the following colors, which are used throughout the visual:

- Primary color

- Primary text color

- Secondary color

- Secondary text color

- Background color

- Font color

These colors are used throughout the whole Key Influencers visual, except for the column charts, which you set as *Drill visual color*:

- Default color

- Reference line color

Data Model

When you play around with the Key Influencers visual (and keep to data modeling best practices), you will sooner or later bump into the error message shown in Figure 3-11: *The analysis is performed at the record level of the 'Product' table. A field in 'Explain by' is not on the 'Product' table or a table related to it by a many-to-one relationship. Try summarizing it.*

Figure 3-21. *Key Influencers is complaining about the record level*

In my case, I provoked this error by putting product category (to be more precise: *EnglishProductCategoryName*) from the *Product* table into *Analyze* (and field *Promotion* from *Reseller Sales*). As you can see in Figure 3-22, the relationship between the three tables is that the *Product* table and the *Promotion* table each have a one-to-many relationship to table *Reseller Sales*. The filter direction for each of those relationships is single, which means that both *Product* and *Promotion* filter *Reseller Sales* but not the other way around. Therefore, a filter from the *Promotion* table does not reach the *Product* table. And that is it, what the visual is complaining about: For the analysis it goes through every single row of the *Product* and looks for an influence from the *Promotion* table. But table *Promotion* does not filter (= influence) table *Product*. Therefore, it can't find any influence.

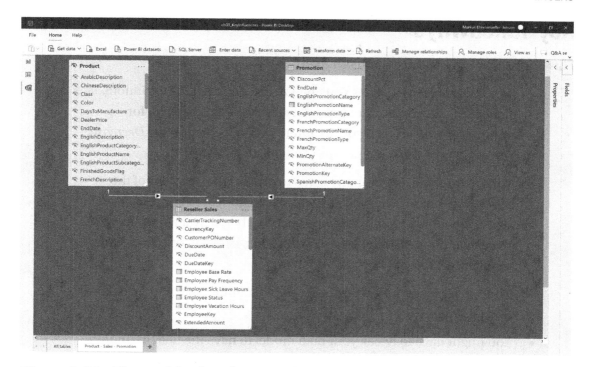

Figure 3-22. *Three tables (Product, Reseller Sales, and Promotion) from the data model in the sample file*

And, no, changing the filter direction of the relationships between those three tables to *Both* will not help here (and should be avoided as much as possible in general, anyways—but that is a story for a different book).

Therefore, I slightly tuned the data model (compared to the examples in the other chapters) to achieve the results shown in this chapter. All the fields I potentially wanted to use in *Analyze* I (re-)created in the *Reseller Sales* table. Then, the Key Influencers visual does the analysis on the level of the table *Reseller Sales*. And *Reseller Sales* is filtered by the *Promotion* table in a one-to-many way. This way, the Key Influencers visual can find out about the influence of *Promotions* on the product category (as it now resides within the *Reseller Sales* table).

There are several ways of moving a column around or (re-)creating it. For this example, I created a new calculated column in table *Reseller Sales* (named *Product Category*) by applying the DAX function RELATED():

```
Product Category = RELATED(ProductCategory[EnglishProductCategoryName])
```

Key Takeaways

This is what you have learned in this chapter:

- Key Influencers visual lets you analyze categorical and continuous fields. For categorical data you select a value from the field, for continuous you select if you are interested in its *Increase* or *Decrease* or you change the *Analysis type* explicitly to *Categorical*.

- You can provide one or many fields to explain the behavior of the preceding field. These fields may be of categorical or continuous data type. Depending on the data type and number of available values, the explanation is performed through a bar, column, or scatter chart visual (see Table 3-1).

- The *Top segments* option looks for helpful combinations of fields to form segments.

- You might have to tweak your data model to have the field in *Analyze* available in the table with the lowest granularity. The columns for the *Explain by* field may be in other tables, connected in such a way that they filter the table where the field for *Analyze* resides.

Categorical and continuous data can be used both to be analyzed or to explain the analyzed field. So, we have the combinations seen in Table 3-1.

Table 3-1. *Combinations of Data Types in Analyze and Explain by and the Type of Chart You Will Get*

	Analyze Categorical	**Analyze Continuous**
***Explain by* Categorical**	Column chart with the selected influencer colored in green and the rest in blue; average as red line	Bar chart with a red line showing the average
***Explain by* Continuous**	Bar chart with a red line showing the average	Scatter chart with a simple regression line in red

CHAPTER 4

Drilling Down and Decomposing Hierarchies

Drilling up and down your hierarchies (e.g., product categories, geography, or calendar) is a very natural way of discovering interesting insights in your data. Many business intelligence tools are capable of such a functionality, and Power BI is no exception. What's special with Power BI is that the actual path you are walking up and down can be tracked in a so-called decomposition tree visual. It's very useful for keeping track of what you already selected and comes in very handy when you want to quickly walk the same path for another part of your hierarchy. But let's start with handling hierarchies the more traditional ways.

Expand and Collapse in a Visual

In Figure 4-1, I build two pairs of similar visuals, for educational purposes only. The real-life purpose of repeating the same information four times is rather limited.

© Markus Ehrenmueller-Jensen 2020
M. Ehrenmueller-Jensen, *Self-Service AI with Power BI Desktop*, https://doi.org/10.1007/978-1-4842-6231-3_4

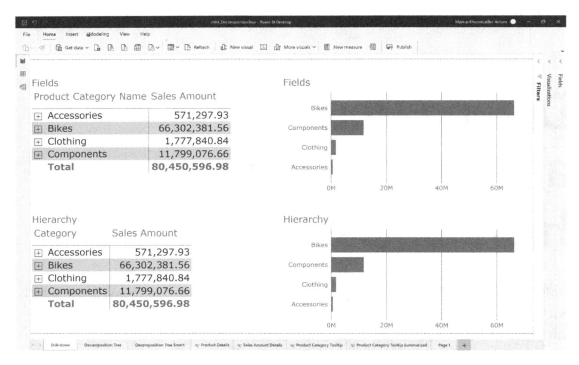

Figure 4-1. *Two tables and two bar charts, with the same content*

We will first concentrate on both table visuals on the left. They make it very apparent that there are more details to discover. They show a plus (+) icon in front of the category name. Clicking the plus icon expands this level and shows the direct sub-level. After clicking the plus icon in front of *Bikes* we get three new lines added to the table, representing mountain, road, and touring bikes (Figure 4-2).

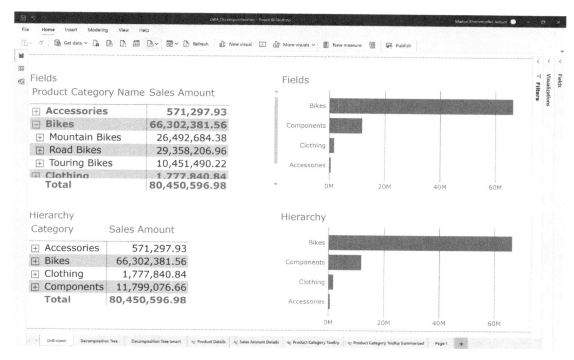

Figure 4-2. *Bikes are expanded*

Clicking the—now appeared—minus (-) icon in front of the *Bikes* category collapses *Bikes*, and the three sub-categories disappear again (and we are back to what we saw in Figure 4-1).

You can achieve the same, and more, when you right-click a category name (e.g., *Bikes*). Figure 4-3 shows all possibilities. Choose *Expand* and *Selection* to expand the category you have clicked on. Choose *Collapse* and *Selection* to collapse the category you have clicked on. Choosing *Entire level* instead of *Selection* does the same as *Expand to next level*, which expands not only the selected category, but all product categories. *Show next Level* does the same, but without showing the product categories. *Expand ➤ All* shows all rows from all levels. With this you basically achieve a detailed report with sums and totals over all available levels.

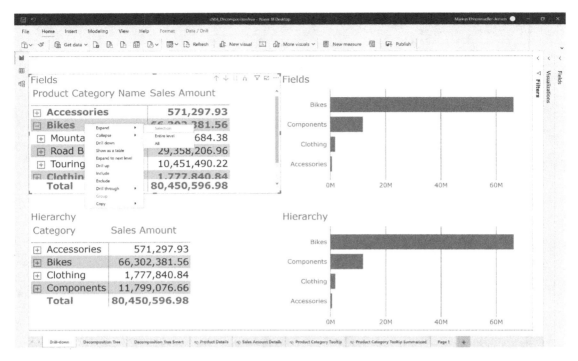

Figure 4-3. *The context menu offers several possibilities to expand or collapse*

The choices in a bar chart (on the right side of the screenshot in Figure 4-3) for expanding and collapsing are fewer. For a right-click we only get the choice of *Expand to next level* and *Show next Level*. You must choose *Drill up* to return to what the visual looked like previously.

All the discussed features are the same for both the visuals with header *Fields* and those with header *Hierarchy*. Read more about the difference in the section "Hierarchies in the Data Model."

Drilling Up and Down in a Visual

A right-click allows you to either *Drill down* or *Drill up* or both, depending on which level of a hierarchy you are currently at (see Figure 4-3).

The difference between drilling and expanding/collapsing is that drilling down expands to the next level while simultaneously limiting all rows to the value in the selected level. If you *Drill down* on *Bikes* in the table visual, all the other product categories but *Bikes* and its sub-categories disappear.

In the bar chart, it is a little bit of a different story, as only the sub-categories are shown, but not the level above (category). The reason for that is that if you add (sub-) totals to a bar or column chart, the individual values might be too small in comparison, leaving the chart rather useless.

As soon as you move the mouse cursor over or click on a visual, its header appears. In the screenshot in Figure 4-4 you see four different kinds of arrows in the upper right corner of the upper table visual, as follows:

- one single arrow pointing up

- one single arrow pointing down

- a double arrow pointing down

- a fork pointing down

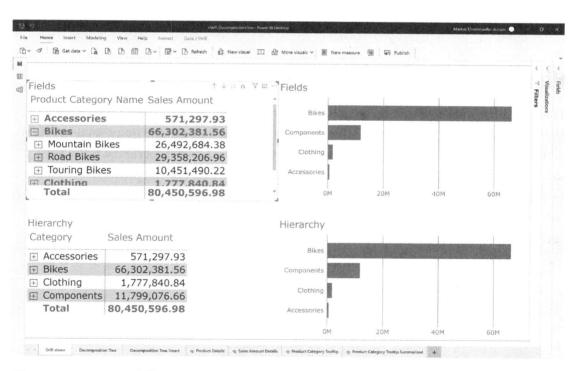

Figure 4-4. *Four different ways of drilling with the header of a visual, represented by four different arrows*

The single arrow pointing up makes it possible to drill up (after you already have drilled down at least one level).

The single arrow pointing down is special in that this icon toggles between two modes. Default behavior in Power BI is that if you click an element in a chart, it will cross-filter the other visuals on the same page. With the mentioned icon (arrow pointing down) you change this behavior to drill-down mode. If drill-down mode is turned on, the single arrow pointing up will be shown in inversed colors: a white arrow inside a black circle. Clicking on a row or a chart element (e.g., a bar) drills down one level (exactly as if you had right-clicked and selected *Drill down*). Drill-down mode stays activated until you turn it off again.

The double arrow pointing down shows the next level (like *Show next level* in the context menu; see earlier section), and the fork expands to the next level (like *Expand to next level* in the context menu).

All the discussed features are the same for both the visuals with header *Fields* and those with header *Hierarchy*. Read more about the difference in the section "Hierarchies in the Data Model."

Hierarchies in the Data Model

The sole difference between the visuals in Figure 4-1 headed *Fields* and those headed *Hierarchy* is that the former were built by adding several fields to the *Rows* section (of the table visual) or the *Axis* section (of the bar chart). The end-user experience is the very same, independent of whether you created the visual one way or the other.

Hierarchies make it easier for the report author though. You can add a set of fields (which are all part of the hierarchy) to a visual at once instead of adding the fields in several steps.

To create your own hierarchy, follow these steps:

1.   Change to *Model* view in Power BI Desktop (Figure 4-5).

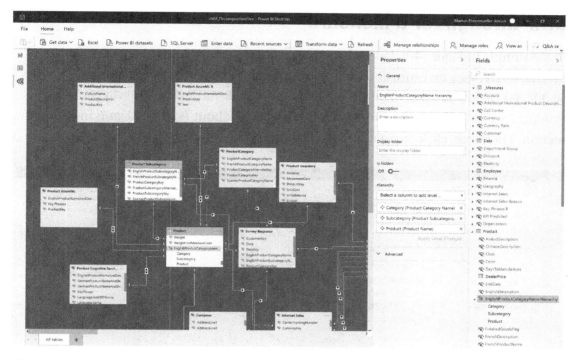

Figure 4-5. *Model view in Power BI Desktop*

2. Right-click on one of the fields (that is not already inside a hierarchy), which should be a member of the hierarchy (either inside of one of the tables in the center or in the field list on the right) and select *Create hierarchy*.

3. Add the other fields in the *General* section of the *Properties* pane by clicking on *Select a column to add level* (Figure 4-5).

4. You can rearrange the order of the fields by dragging and dropping them.

5. Don't forget to click on *Apply Level Changes* or you must re-do all of the steps (Figure 4-5).

Drill-through

Drill-through is available for a measure, for a column, or to a different report.

Drill-through for a Measure

In the sample in Figure 4-6, I already prepared the necessary steps to enable a drill-through (for a measure and for a column). Right-click on the sales amount 66,302,381.56 in line *Bikes* in one of the two tables. Drill-through offers two choices: *Product Details* and *Sales Amount Details*. We will talk about *Product Details* in the next section. Therefore, click on *Sales Amount Details* now.

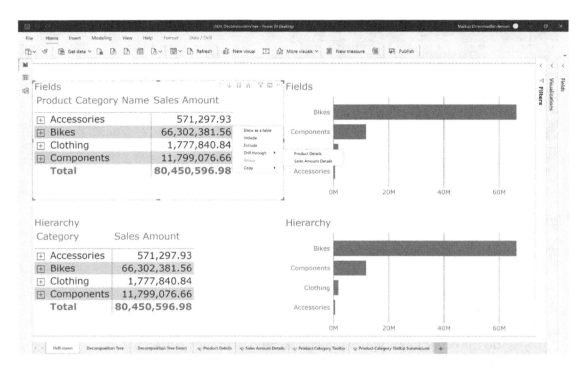

Figure 4-6. *Drill-through is available in the context menu for Sales Amount in this example*

Power BI will jump over to a page with the exact same name (*Sales Amount Details*) and show you the sum of sales amount for bikes on the top left (66.30M) and a list of orders in a table visual (including *Order Number, Line Number, Product, Sales Amount,* and *Order Quantity*).

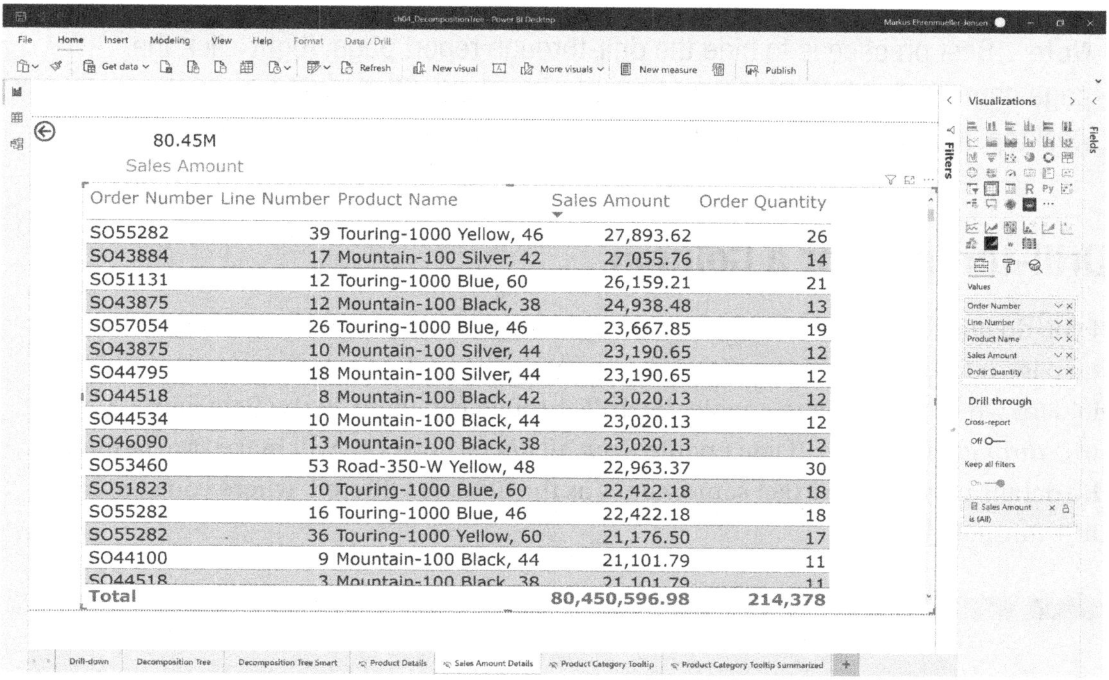

Figure 4-7. *Sales amount details*

With the button on the top left of the report (circle with an arrow pointing to the left) you can return to the previous report. In Power BI Desktop you have to hold the control key while you click the button to activate its action, as a simple click will just select the object. In Power BI Service an ordinary click is enough to activate the action. If you select a different sales amount number in the table visual in Figure 4-6 as starting point, a different product category will be pre-selected in the *Sales Amount Details* report.

Sales Amount Details is a page I created and made after my own imagination, with what I thought could be interesting details for the sales amount. To make a report page available as a drill-through report you must add a measure into the *Drill-through* section in the *Visualizations* pane (see Figure 4-7). That's it. From now on this report will be listed in the *Drill-through* context menu for this measure—anywhere in the current report file. If the measure is currently filtered (like the 66M are the sales amount for bikes only), this filter will automatically be brought over to the details report.

Note   Best practice is to hide the drill-through report page. Right-click the page name and select *Hide Page*. It is then marked with an eye icon in Power BI Desktop. In Power BI Service, the page is indeed hidden.

Drill-through for a Column

If you do not click on *Sales Amount Details* in Figure 4-6, but rather on *Product Details*, the page named *Product Details* will be opened. The trick here is very similar: add the field for which you want to enable the drill-through capability in the field name *Add drill-through fields here*. If you enable *Keep all filters*, Power BI will make sure to filter the details page for the exact same values as the field was filtered, where you started the drill-through from.

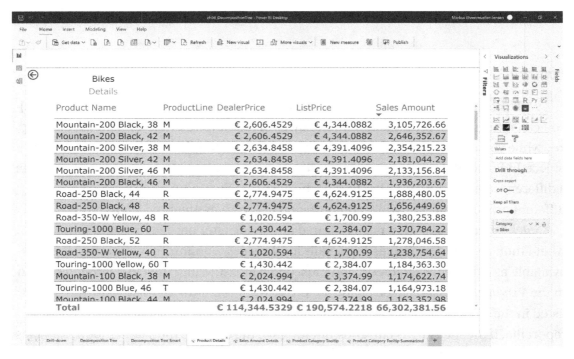

Figure 4-8. *Product details*

Note   Best practice is to hide the drill-through report page. Right-click the page name and select *Hide Page*. It is then marked with an eye icon in Power BI Desktop. In Power BI Service, the page is indeed hidden.

Drill-through to a Different Report

In the Power BI Service, you can enable drilling through to a page inside a different report. The feature only works for reports that are published to the same workspace in the Power BI Service. It does not work from Power BI Desktop.

It is crucial that both the table and the column are named identically in the model of the two reports involved (the comparison is case sensitive!). Then, you must explicitly change options in both files.

In the file you want to start the drill-through experience from (the source report) you have to enable the feature by enabling option *Allow visuals in this report to use drill-through targets from other reports* in *File ➤ Options ➤ Current File ➤ Report settings* and republish the report. Alternatively, you can enable the setting with the same name in the workspace directly in the service.

In the target report, enable *Cross-report* in the *Drill-through* section of the *Format* options of the page, to which you want to drill-through.

Find out more about this feature here: `https://docs.microsoft.com/en-us/power-bi/desktop-cross-report-drill-through`.

Tooltip

All these features need you to click at least twice before you get to the details. A tooltip only needs you to move the mouse cursor over a value shown in a visual. The default behavior of the tooltip in Power BI Desktop is that the values of the axis and the numeric value(s) of the chart are shown in numbers (Figure 4-9).

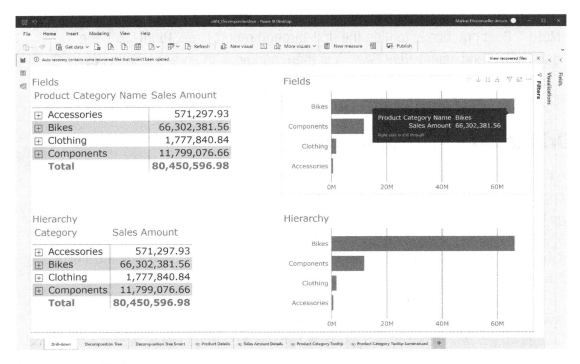

Figure 4-9. *Default tooltip*

In Figure 4-10, you see a tooltip I explicitly created. It shows the name of the product category, the sum of sales amount, the average of sales amount (per sale), and a bar chart for all of the sub-categories.

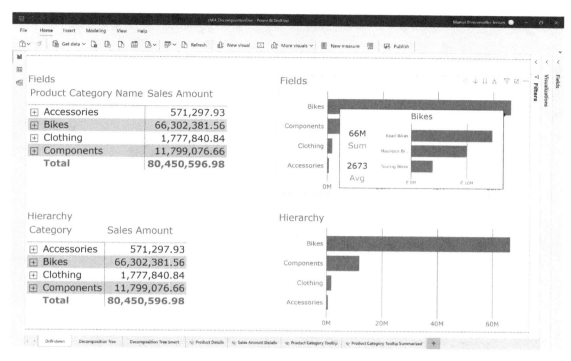

Figure 4-10. *Custom tooltip*

To (re-)build the example, follow these steps:

1. Create a new report page.

2. In *Format* options for the page (see Figure 4-11) a) enable *Tooltip* and b) change *Page size* to either *Tooltip* or *Custom* (otherwise the tooltip would be as large as the report itself).

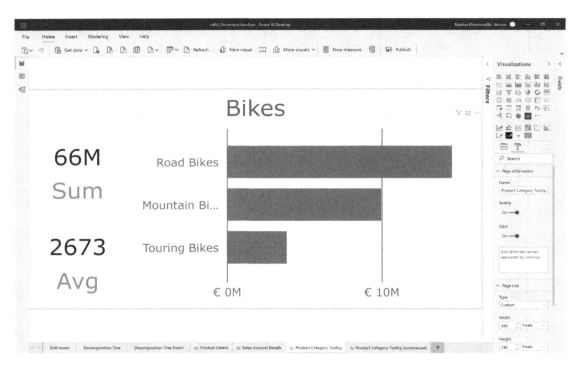

Figure 4-11. *Turn on the Tooltip setting and make the page size of the tooltip report smaller than the original report*

3. Add the fields and visuals you want to be part of your custom tooltip.

4. In the *Fields options*, add a categorical field for which you want this tooltip to be available. This is very similar to the drill-through functionality, explained further earlier in this chapter.

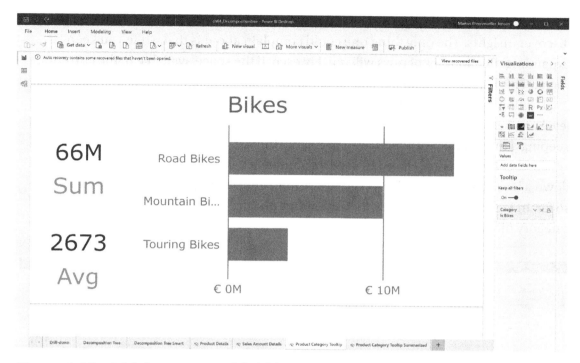

Figure 4-12. *Add the categorical field for which the tooltip should be enabled in the Tooltip section*

Note   Best practice is to hide the tooltip report page. Right-click the page name and select *Hide Page*. In Power BI Service, the page is indeed hidden.

Decomposition Tree (traditional)

We are now mentally jumping back to the expand/collapse and drill-down/-up features described earlier. Remember: Expanding/collapsing adds or removes rows from a lower level of a hierarchy to the current visual (e.g., adding the sub-categories to the already visible categories). Drilling basically either changes the current level to a different one (e.g., showing the sub-categories instead of the categories) or adds a filter to the visual so as to only show a certain part of the hierarchy tree (e.g., showing the categories and sub-categories for only the selected category).

I like these features because they allow you to interactively explore the data and discover insights. The only thing one could complain about is that while you keep expanding, the number of rows will quickly exceed the space available in the current visual. This leaves you with the task of scrolling up and down.

Drilling up and down the hierarchies minimizes this effect a little, as only a certain level or a certain part of the hierarchy tree is visual. But this comes with the challenge of keeping a mental map of which level you came from so you do not get lost after a while.

All these caveats are solved by the Decomposition Tree visual. It allows you to drill down a level (not all rows of all levels are visible) and visually see the path you walked so far. In Figure 4-13, I let a decomposition tree analyze *Sales Amount* over the product category hierarchy.

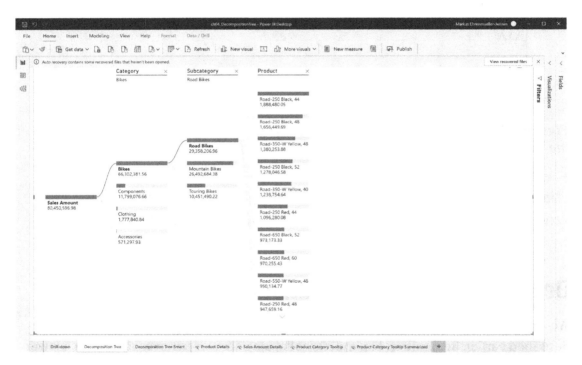

Figure 4-13. *Decomposition tree*

The visual starts with a total over *Sales Amount* on the very left, which is split by *Category* (*Bikes, Components, Clothing*, and *Accessories*) just right of it. The categories are sorted by *Sales Amount*—you can't change the order or the field.

To change the path, you just click on another element of the decomposition tree. If you click, for example, on *Components*, you get the sub-categories of *Components* listed. Which categories (e.g., product category, product sub-category, and product name) are available in the tree is controlled by the fields in *Explain by.*

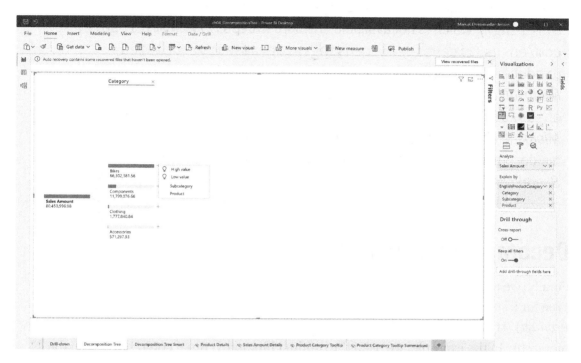

Figure 4-14. *The available levels are controlled by Explain by. You can change the order of the levels by either adding new ones (+) or removing unwanted ones (x)*

The order of the categories (e.g., category on first, sub-category on second, and product on third level) is controlled when you open a level the first time via the plus icon. If you are not happy with the levels, you can remove them by clicking on the x icon, just beside the header of each level at the very top of the decomposition tree, and adding a different one via the plus icon (see Figure 4-14).

Note   Decomposition trees are nothing new but have a long history of up's and down's. They were already available in *ProClarity*, a tool bought by Microsoft back in 2006 (`https://news.microsoft.com/2006/04/03/microsoft-agrees-to-acquire-proclarity-enhancing-business-intelligence-offering/`), which is not available anymore (`https://www.microsoft.com/en-us/licensing/licensing-programs/isvr-deleted-products-proclarity`). The visual appeared later in the now deprecated *Microsoft PerformancePoint Server 2007* (`https://docs.microsoft.com/en-us/office365/enterprise/pps-2007-end-of-support`) and in *Microsoft SharePoint's Performance Point Services*. With the end of *Silverlight* the decomposition tree also was removed from *PerformancePoint Services* in *SharePoint 2019*.

Decomposition Tree (smart)

Finally, we are back in the field of AI. In the previous section, we asked Decomposition Tree for a specific level when opening another section (e.g., first *Sub-category*, then *Product*). This is useful in scenarios where you are exploring a so-called natural hierarchy. In the example, we used the product hierarchy. But it could have been the sales territories (e.g., groups, country, and region) or time (e.g., year, quarter, month, day, shift, etc.) or an organizational hierarchy (e.g., departments and employees) as well. I'm sure you can find many different hierarchies in your organization.

Decomposition Tree can select the field for the next level automatically by considering which is the most influential one on the measure you are analyzing. This might remind you of the capabilities of the Insights feature (Chapter 2) and the Key Influencer Visual (Chapter 3). Decomposition Tree is indeed the third smart visual Power BI came up with. (All three visuals have a lightbulb in their icon.)

The smart way of using Decomposition Tree is therefore to put possible influencers into the *Explain by* field (Figure 4-15).

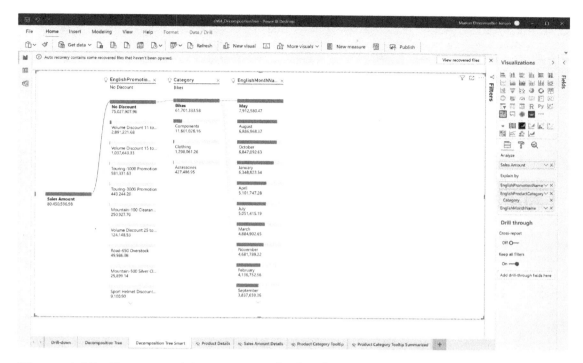

Figure 4-15. *Decomposition Tree can find influencers*

When using the visual this way, you do not ask for a specific level or field, but rather click on either *High value* or *Low value* (see Figure 4-14). Both options are, again, marked with a lightbulb to make it clear that there is a machine learning model working behind the scenes to find significant influencers. A lightbulb appears on the very top of each section in the Decomposition Tree, as well. This indicates that those categories are not static, but rather might change depending on what you select on the upper levels.

This is what we see in Figure 4-15: The biggest influencer on *Sales Amount* for *EnglishPromotionName* (level 1 of the decomposition tree) with *No Discount* in field *Category* (level 2) for *Bikes* is month (level 3) *May*. If you click on promotion *Mountain-100 Clearance Sale* in the first level, the structure of the decomposition tree will change. It will first show the month (*November*) on level 2 and the category (*Bikes*) on level 3 of the visual, as you can see in Figure 4-16.

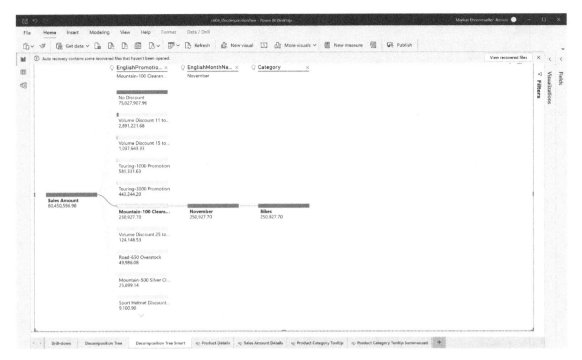

Figure 4-16. *Decomposition Tree did smartly change the fields in the levels for a different selection to find influencers*

Key Takeaways

This is what you have learned in this chapter:

- Hovering the mouse cursor over a number represented in a chart will show details for this number. You can customize this tooltip by creating your own detailed view.

- Hierarchies exist in almost all data (categorical levels, geography, time, etc.). You can explore hierarchies in most of the visuals by adding more than one categorical field. If you need the same combination of fields repeatedly, you might create a hierarchy in the data model for that combination for convenience reasons.

- Expanding a hierarchy will add rows of a "lower" level, while collapsing will remove them again.

- Drilling down a hierarchy will limit the shown rows to the current selection and simultaneously expand to the next level. Drilling up reverses this step.

- Drilling through a measure, a column, or a different report will open a different report (page) and set the filter to the current selection.

- Decomposition Tree visually shows the path you drill down. You can explicitly ask for a certain sub-level or let the visual find a level with the biggest influence on the number on which you want to drill down.

CHAPTER 5

Adding Smart Visualizations

We have already seen some smart visualizations in the previous chapters (Q&A, Key
Influencers, and Decomposition Tree). Besides those standard visuals, you have access
to so-called custom visuals from a marketplace (called *AppSource*). The marketplace
is maintained by Microsoft and is filled with interesting additional visualizations built
by several different companies (Microsoft and others). Most of them are free to use,
but some will ask you for a fee before you can use their full capabilities (such Power BI
visuals are hinted with *May require additional purchase*). In this chapter, we will look at
a selection of visualizations offered by Microsoft under the section "Advanced Analytics."
All of the following are free to use:

- Time Series Forecasting Chart

- Forecasting with ARIMA

- Forecasting TBATS

- Time Series Decomposition Chart

- Spline Chart

- Clustering

- Clustering with Outliers

- Outliers Detection

- Correlation Plot

- Decision Tree Chart

- Word Cloud

© Markus Ehrenmueller-Jensen 2020
M. Ehrenmueller-Jensen, *Self-Service AI with Power BI Desktop*, https://doi.org/10.1007/978-1-4842-6231-3_5

Most of those custom visualizations are powered by R behind the scenes. These visuals enable you to use the power of R (in its packages) without writing a single line of code (so-called no-code solution). You can replicate all those visuals with your own R script—and I will show you some of them later in this book, in Chapter 9, "Executing R and Python Visualizations."

But before we explore these new visuals, I will show you how to add the following smart information to a line chart (and later to a scatter chart):

- Trendline

- Trendline in DAX

- Forecast

Trendline

Reseller sales amount in the *Adventure Works* database is quite volatile over time (column *'Date'[Date]*): There are many ups and downs over the months, as you can see in Figure 5-1.

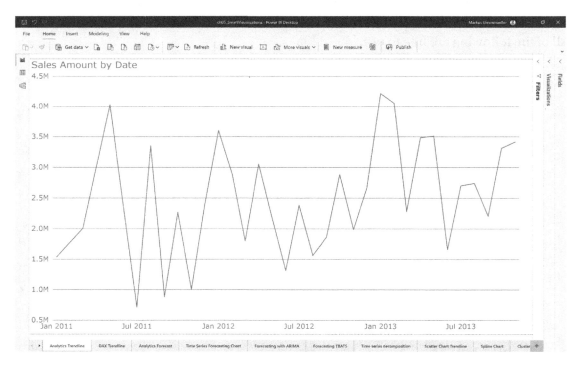

Figure 5-1. *Sales Amount by Date is volatile*

Over the whole timeframe, it looks like there is an upward trend though. With Power BI, we don't have to depend on a gut feeling—we can add a trendline with just a few mouse clicks. Similar to the constant line we added in Chapter 2 ("The Insights Feature") to the *100% Stacked bar chart*, we have the choice of different analytic lines we can add to a line chart, as follows:

- Trendline

- Min

- Max

- Constant

- Forecast

In Figure 5-2, you can see the added trendline (you will find an example with the Forecast line further on).

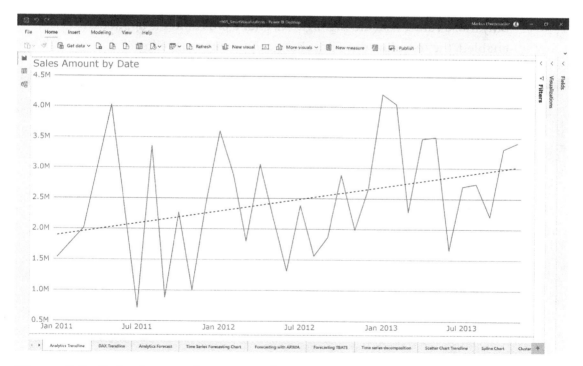

Figure 5-2. *Sales Amount by Date with a trendline added*

The options to influence the trendline are only a few (Figure 5-3), as follows:

- You can change the name of the trendline by double-clicking the small box with "Trend line 1" in it. You can delete the trendline by clicking on the x on the right.

- *Color*

- *Transparency* can be a percentage between 0 and 100.

- *Style*: *Dashed*, *Solid*, or *Dotted*

- *Combine Series*: To try this feature you must first put a field (e.g., *Product Category*) into the *Legend*. If *Combine Series* is turned off, you get one trendline per line in the line chart, representing a value of the legend. If turned on, only a single combined trendline is drawn.

- *Use Highlight Values*: Does only show an effect in clustered column charts (not in a line chart). If disabled, the trendline is calculated for only those values that are highlighted, after a cross-filter is applied. If enabled, the trendline does not change upon any cross-filter.

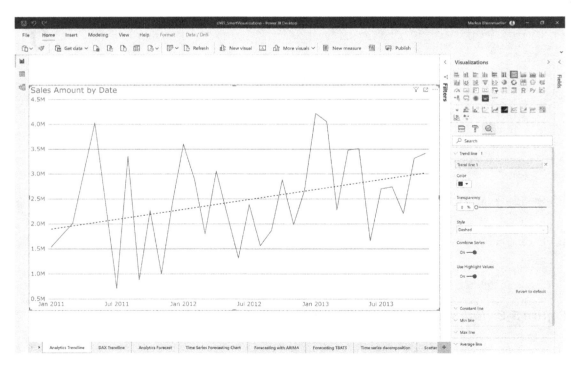

Figure 5-3. *The options to influence a trendline*

As a matter of fact, we cannot change the way the trendline is calculated. The calculation is implemented in the code of the line chart, and it does not surface any parameters to influence it. But we can calculate the trendline ourselves, which you will learn next.

Trendline in DAX

The trendline available out of the box is calculated as a simple linear regression. In this section, we will reconstruct the built-in trendline as a DAX measure. DAX (short for Data Analytic Expression) is the language to create calculations in Power BI. This link is a good starting point, if you are new to DAX: `https://docs.microsoft.com/en-us/power-bi/transform-model/desktop-quickstartlearn-dax-basics`. The only reason you would reconstruct a built-in functionality is because having the calculation as a measure at hand gives you the flexibility to change it for your needs. You will find some applications at the end of this section. I hope I do not scare you away by plotting the formulas here:

$$y = \text{Intercept} + \text{Slope} * x$$

With *Intercept* as the point where the line crosses the y-axis and with *Slope* as the amount of how much (or less) *y* increases (or decreases in the matter of a negative slope) for every increase of *x*. Both can be estimated via the ordinary least square method (bear with me!):

$$\text{Slope} = \frac{\text{Cov}(x,y)}{\text{Var}(x)} = \frac{N * \sum(x*y) - \sum x * \sum y}{N * \sum x^2 - (\sum x)^2}$$

$$\text{Intercept} = \bar{y} - \text{Slope} * \bar{x}$$

Exactly this formula is implemented in the following example in DAX, based on a blog post by Daniil Maslyuk (`https://xxlbi.com/blog/simple-linear-regression-in-dax/`). I will explain it part by part (in didactical order, not in order of execution); you will find the full code example at the end.

This code creates two variables in DAX for `Slope` and `Intercept`:

```
// Slope & Intercept
 VAR Slope =
    DIVIDE (
        N * SumOfXTimesY - SumOfX * SumOfY,
        N * SumOfXSquared - SumOfX ^ 2
    )
```

```
VAR Intercept =
    AverageOfY - Slope * AverageOfX
```

The preceding formula uses other variables (like N or SumOfXTimesY), which are calculated here:

```
// Parts of formula
VAR N =
    COUNTROWS ( Actual )
VAR SumOfXTimesY =
    SUMX ( Actual, Actual[X] * Actual[Y] )
VAR SumOfX =
    SUMX ( Actual, Actual[X] )
VAR SumOfY =
    SUMX ( Actual, Actual[Y] )
VAR SumOfXSquared =
    SUMX ( Actual, Actual[X] ^ 2 )
VAR AverageOfY =
    AVERAGEX ( Actual, Actual[Y] )
VAR AverageOfX =
    AVERAGEX ( Actual, Actual[X] )
```

Actual is another variable, which contains a table of values for measuring [Sales Amount] per available 'Date'[Date]:

```
// Actual values on which the regression line is based
VAR Actual =
    FILTER (
        SELECTCOLUMNS (
            ALLSELECTED ( 'Date'[Date] ),
            "Actual[X]", 'Date'[Date],
            "Actual[Y]", [Sales Amount]
        ),
        AND (
            NOT ( ISBLANK ( Actual[X] ) ),
            NOT ( ISBLANK ( Actual[Y] ) )
        )
    )
```

Finally, the calculation result is returned from the following expression. The `FILTER` makes sure that the trendline is only shown for dates where there is at least one row in table *Reseller Sales* available (which is the behavior of the trendline in the *Analytics* pane).

```
// Return values for trendline
 VAR RET =
    CALCULATE (
        SUMX (
        DISTINCT ( 'Date'[Date] ),
        Intercept + Slope * 'Date'[Date]
        ),
        FILTER (
            DISTINCT ( 'Date'[Date] ),
            CALCULATE ( COUNTROWS ( 'Reseller Sales' ) ) > 0
        )
    )
```

To create the measure `Simple Linear Regression`, select *Modeling* ➤ *New measure* in the ribbon and paste the following code:

```
Simple Linear Regression =
/* Based on "Simple linear regression in DAX" by Daniil Maslyuk
 * https://xxlbi.com/blog/simple-linear-regression-in-dax/
 */

// Actual values on which the regression line is based
VAR Actual =
    FILTER (
        SELECTCOLUMNS (
            ALLSELECTED ( 'Date'[Date] ),
            "Actual[X]", 'Date'[Date],
            "Actual[Y]", [Sales Amount]
        ),
        AND (
            NOT ( ISBLANK ( Actual[X] ) ),
            NOT ( ISBLANK ( Actual[Y] ) )
        )
    )
```

```
// Parts of formula
VAR N =
    COUNTROWS ( Actual )
VAR SumOfXTimesY =
    SUMX ( Actual, Actual[X] * Actual[Y] )
VAR SumOfX =
    SUMX ( Actual, Actual[X] )
VAR SumOfY =
    SUMX ( Actual, Actual[Y] )
VAR SumOfXSquared =
    SUMX ( Actual, Actual[X] ^ 2 )
VAR AverageOfY =
    AVERAGEX ( Actual, Actual[Y] )
VAR AverageOfX =
    AVERAGEX ( Actual, Actual[X] )

// Slope & Intercept
VAR Slope =
    DIVIDE (
        N * SumOfXTimesY - SumOfX * SumOfY,
        N * SumOfXSquared - SumOfX ^ 2
    )
VAR Intercept =
    AverageOfY - Slope * AverageOfX

// Return values for trendline
VAR RET =
    CALCULATE (
        SUMX (
            DISTINCT ( 'Date'[Date] ),
            Intercept + Slope * 'Date'[Date]
        ),
```

```
        FILTER (
            DISTINCT ( 'Date'[Date] ),
            CALCULATE ( COUNTROWS ( 'Reseller Sales' ) ) > 0
        )
    )
RETURN
    RET
```

In Figure 5-4, I removed the trendline from the *Analytic* pane and added the newly created measure *Simple Linear Regression* to the fields.

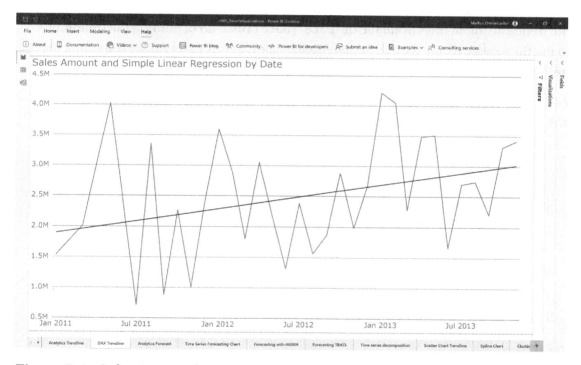

Figure 5-4. *Sales Amount by Date with a trendline calculated in DAX added*

The result is identical to Figure 5-3 (except that the trendline is now solid and not dashed). "Why should I then bother with this lengthy DAX code?" you might think. If you need the trendline exactly like this, the answer will be not to bother with the lengthy DAX code. In case you need to change the calculation of the trendline, this lengthy DAX code is your friend, as you will see in next examples. We will adopt the trendline in three steps:

- Prolong the trendline over dates where no actual value is available ("into the future").

- Shorten the trendline to start in year 2013.

- Change the calculation of the trendline to be based on data from 2013 onward only, as more current data could lead to a more realistic trendline.

First, we remove the FILTER expression in the variable RET to deliver the values for the trendline for all rows available in 'Date'[Date] (instead of limiting the trendline to the time range where we had actual reseller sales; in this case, we are extending the trendline into future years; I marked the change in bold font):

```
// Return values for trendline
VAR RET =
    CALCULATE (
        SUMX (
            DISTINCT ( 'Date'[Date] ),
            Intercept + Slope * 'Date'[Date]
            // FILTER removed
        )
    )
```

Then, we reintroduce a filter expression to variable RET (printed in bold font). This will make sure to only show values for the trendline from January 1, 2013, onward:

```
VAR RET =
    CALCULATE (
        SUMX (
            DISTINCT ( 'Date'[Date] ),
            Intercept + Slope * 'Date'[Date]
        ),
        FILTER (
            DISTINCT ( 'Date'[Date] ),
```

```
            'Date'[Date] >= DATE(2013, 01, 01)
        )
    )
```

Finally, we exchange the ALLSELECTED in variable Actual with a similar filter as in RET to make sure that only values from 2013 onward are used for the calculation of the trendline (again the change to the original script is printed in bold font):

```
// Actual values on which the regression line is based
VAR Actual =
    FILTER (
        SELECTCOLUMNS (
            FILTER (
                ALLSELECTED ( 'Date'[Date] ),
                'Date'[Date] >= DATE(2013, 01, 01)
            ),
            "Actual[X]", 'Date'[Date],
            "Actual[Y]", [Sales Amount]
        ),
        AND (
            NOT ( ISBLANK ( Actual[X] ) ),
            NOT ( ISBLANK ( Actual[Y] ) )
        )
    )
```

The truth is that from the values in year 2013 alone we derive a negative trend (see report *DAX Trendline* as shown in Figure 5-5). Hopefully, we can find measurements to turn this downward trend around and see *Adventure Works* into a prosperous future.

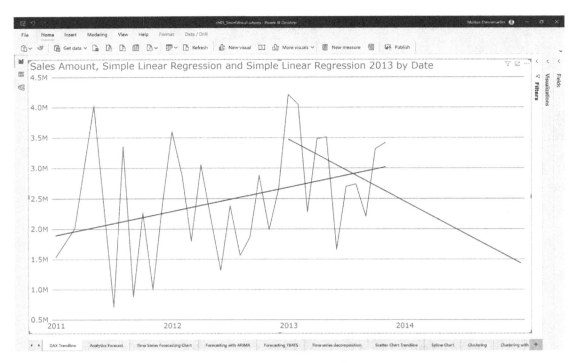

Figure 5-5. *Sales Amount by Date with two different trendlines calculated in DAX*

Find the full code for measure `Simple Linear Regression 2013` here (with the changes made in bold font):

```
Simple Linear Regression 2013 =
/* Based on "Simple linear regression in DAX" by Daniil Maslyuk
 * https://xxlbi.com/blog/simple-linear-regression-in-dax/
 */

// Actual values on which the regression line is based
VAR Actual =
    FILTER (
        SELECTCOLUMNS (
            FILTER (
                ALLSELECTED ( 'Date'[Date] ),
                'Date'[Date] >= DATE(2013, 01, 01)
            ),
            "Actual[X]", 'Date'[Date],
            "Actual[Y]", [Sales Amount]
        ),
```

```
        AND (
            NOT ( ISBLANK ( Actual[X] ) ),
            NOT ( ISBLANK ( Actual[Y] ) )
        )
    )
// Parts of formula
VAR N =
    COUNTROWS ( Actual )
VAR SumOfXTimesY =
    SUMX ( Actual, Actual[X] * Actual[Y] )
VAR SumOfX =
    SUMX ( Actual, Actual[X] )
VAR SumOfY =
    SUMX ( Actual, Actual[Y] )
VAR SumOfXSquared =
    SUMX ( Actual, Actual[X] ^ 2 )
VAR AverageOfY =
    AVERAGEX ( Actual, Actual[Y] )
VAR AverageOfX =
    AVERAGEX ( Actual, Actual[X] )

// Slope & Intercept
VAR Slope =
    DIVIDE (
        N * SumOfXTimesY - SumOfX * SumOfY,
        N * SumOfXSquared - SumOfX ^ 2
    )
VAR Intercept =
    AverageOfY - Slope * AverageOfX

// Return values for trendline
VAR RET =
    CALCULATE (
        SUMX (
            DISTINCT ( 'Date'[Date] ),
            Intercept + Slope * 'Date'[Date]
        ),
```

```
        FILTER (
            DISTINCT ( 'Date'[Date] ),
            'Date'[Date] >= DATE(2013, 01, 01)
        )
    )

RETURN
    RET
Simple Linear Regression =
/* Based on "Simple linear regression in DAX" by Daniil Maslyuk
 * https://xxlbi.com/blog/simple-linear-regression-in-dax/
 */

// Actual values on which the regression line is based
VAR Actual =
    FILTER (
        SELECTCOLUMNS (
//          FILTER ( ALLSELECTED ( 'Date'[Date] ); 'Date'[Date] >=
DATE(2013; 01; 01) );
            ALLSELECTED ( 'Date'[Date] ),
            "Actual[X]", 'Date'[Date],
            "Actual[Y]", [Sales Amount]
        ),
        AND (
            NOT ( ISBLANK ( Actual[X] ) ),
            NOT ( ISBLANK ( Actual[Y] ) )
        )
    )

// Parts of formula
VAR N =
    COUNTROWS ( Actual )
VAR SumOfX =
    SUMX ( Actual, Actual[X] )
```

```
VAR SumOfXSquared =
    SUMX ( Actual, Actual[X] ^ 2 )
VAR SumOfY =
    SUMX ( Actual, Actual[Y] )
VAR SumOfXTimesY =
    SUMX ( Actual, Actual[X] * Actual[Y] )
VAR AverageOfX =
    AVERAGEX ( Actual, Actual[X] )
VAR AverageOfY =
    AVERAGEX ( Actual, Actual[Y] )

// Slope & Intercept
VAR Slope =
    DIVIDE (
        N * SumOfXTimesY - SumOfX * SumOfY,
        N * SumOfXSquared - SumOfX ^ 2
    )
VAR Intercept =
    AverageOfY - Slope * AverageOfX

// Return values for trendline
VAR RET =
    CALCULATE (
        SUMX (
            DISTINCT ( 'Date'[Date] ),
            Intercept + Slope * 'Date'[Date]
        ) ,
        FILTER (
            DISTINCT ( 'Date'[Date] ),
            CALCULATE ( COUNTROWS ( 'Reseller Sales' ) ) > 0
//            'Date'[Date] >= DATE(2013; 01; 01) )
        )
    )

RETURN
    RET
```

Forecast

The name of the formula in the previous section says it all: simple regression line. This is a simple method to visualize the trend in your data. If the data is as volatile as the reseller sales amount in *Adventure Works*, a trendline is not very good at giving us a prediction. Here, the *Forecast* in the *Analytics* pane comes into play, which you can see in Figure 5-6.

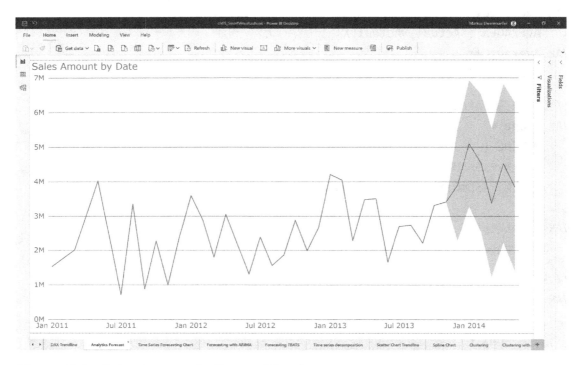

Figure 5-6. *Sales Amount by Date with a forecast*

I replaced the trendline with a line for the forecast for the next six months, including a band for a 95 percent confidence interval. This means that the true value lies inside of the blueish area with a 95 percent probability. This area is rather wide because of the high volatility of the data in the previous months.

It is important to set the seasonality to the right amount (Figure 5-7). I used a value of 12 because my actual data consists of monthly values, and I expect that there is a seasonal pattern inside twelve months, which has the potential to repeat again for a different portion of twelve months. If you look closely, then you can recognize that the

M-shaped pattern for the forecast for months from November 2013 until May 2014 looks like the pattern of both the current line for months November 2011 until May 2012 and November 2012 until May 2013.

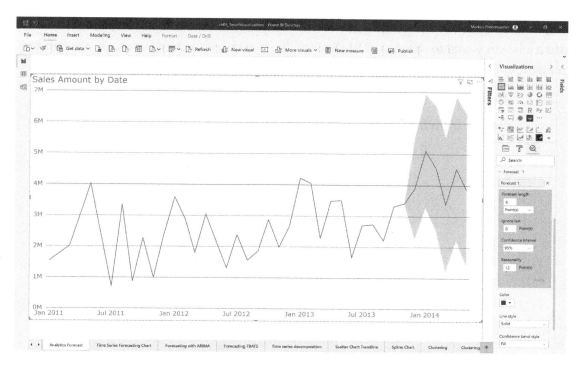

Figure 5-7. *We've got different options to influence the forecast*

If you need more flexibility with the algorithm and parameters for the forecast, you could either replicate this functionality in DAX, like we did for the trendline, or check out Chapter 9, "Executing R and Python Visualizations," where I show you how to invoke machine learning models in scripts in R and Python. Alternatively, we now come to smart custom visualizations, which give a few more options than the *Forecast* in the *Analytics* pane, but don't force you to write any code.

Adding a Custom Visualization

Beyond the standard visuals that are available when you create a new Power BI file, you can add custom visuals by clicking on "..." at the end of the list of standard visuals (Figure 5-8), or in the ribbon going to *Insert* ➤ *More Visuals*. There are two methods to add a custom visual to the Power BI Desktop file you currently have open, as you can see in Figure 5-8:

- Import from AppSource (from the marketplace in the cloud).

- Import from file (from a local PBIVIZ file).

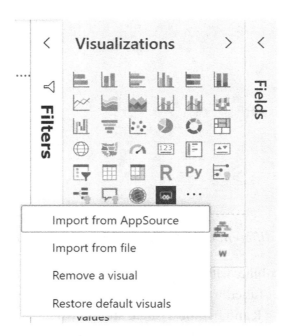

Figure 5-8. *Adding a custom visualization*

Importing from a file will get you a warning that you only should import a custom visual if you trust its author and source (Figure 5-9).

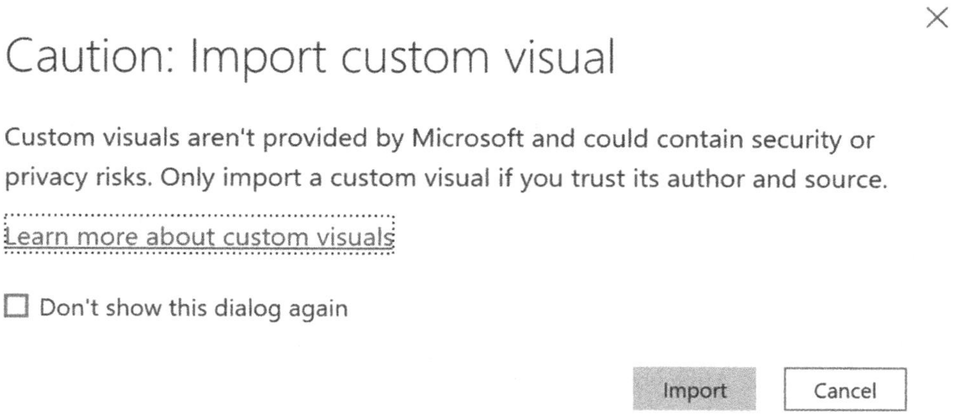

×

Caution: Import custom visual

Custom visuals aren't provided by Microsoft and could contain security or
privacy risks. Only import a custom visual if you trust its author and source.

Learn more about custom visuals

☐ Don't show this dialog again

Import   Cancel

Figure 5-9. *Be cautious when importing custom visuals from a file*

Importing from AppSource will open a list of Power BI visuals available from either
the *Marketplace* or *My Organization* (Figure 5-10).

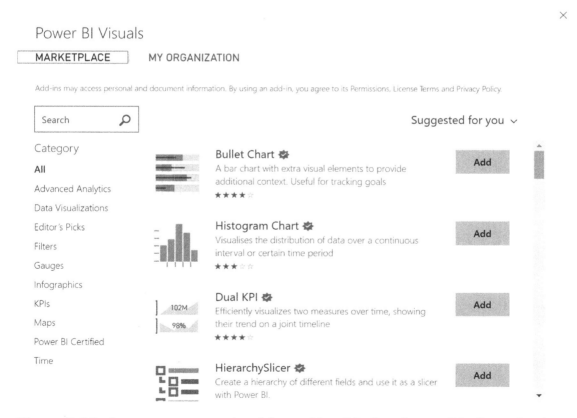

Figure 5-10. *Import a custom visual from either Marketplace or My Organization*

You can browse through the list of add-ins either with the scrollbar handle on the right, by *Category* on the left, or by entering a term in the *Search* box on the upper left. Change the sort order of the list by toggling *Suggested for you* to *Rating* or *Name*.

If you click on the name of the visual, you will see a more detailed description of the custom visual. There, you will get information on which kinds of algorithms are used (or which R packages are used). Some of the custom visuals described in the following sections are open sourced, and you will find a link to GitHub in the description. Most of the custom visuals I discuss in this book require you to install R first. Please refer to either `https://www.r-project.org/` or `https://mran.microsoft.com/open` to download and install R. After you have successfully added custom visuals, please save the current file (to not to lose any changes you might have done so far), close Power BI, and open it up again. In some cases, you are prompted to install required R packages (Figure 5-11). If you do not do so and the visual needs a certain R package, the visual will fail and you will get an error message about a missing package. If you are curious about what is contained in a specific R package, you can learn more at `https://cran.r-project.org/web/packages/`.

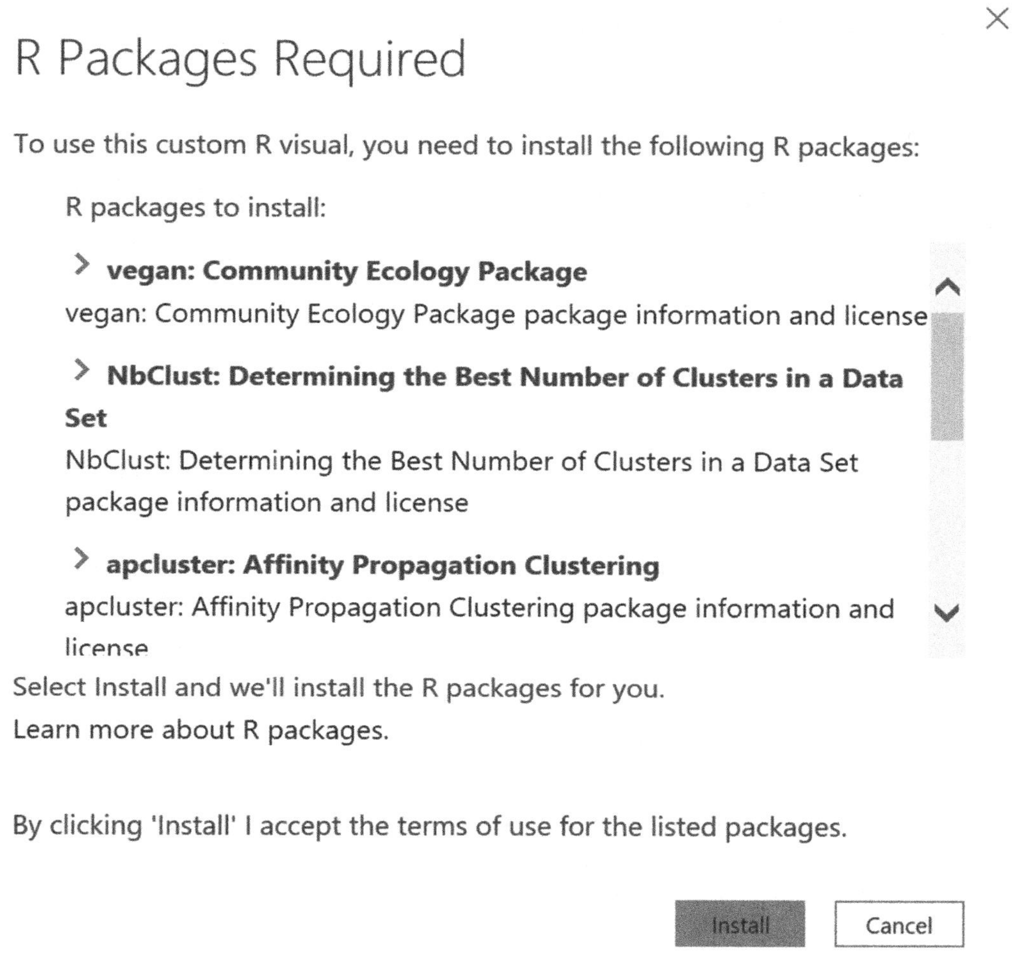

✕

R Packages Required

To use this custom R visual, you need to install the following R packages:

R packages to install:

> **vegan: Community Ecology Package**

vegan: Community Ecology Package package information and license

> **NbClust: Determining the Best Number of Clusters in a Data Set**

NbClust: Determining the Best Number of Clusters in a Data Set package information and license

> **apcluster: Affinity Propagation Clustering**

apcluster: Affinity Propagation Clustering package information and license

Select Install and we'll install the R packages for you.

Learn more about R packages.

By clicking 'Install' I accept the terms of use for the listed packages.

Install   Cancel

Figure 5-11. *Some custom visuals require you to have installed R and will prompt you to install additional R packages*

Attention   Custom visualizations contain code written by either Microsoft or third parties. If you want to be sure that the visual definitely does not access external services or resources (which could potentially leak the data you are visualizing), stick to custom visuals specially marked with a checkmark, which stands for *The visual is certified by Power BI.* All visuals listed in Figure 5-10 show this blue checkmark, just after the visual's name. In the following sections, I stick to certified visuals published by Microsoft.

Time Series Forecasting Chart

You can see an example of a time series forecasting chart in Figure 5-12, which uses R package *forecast*.

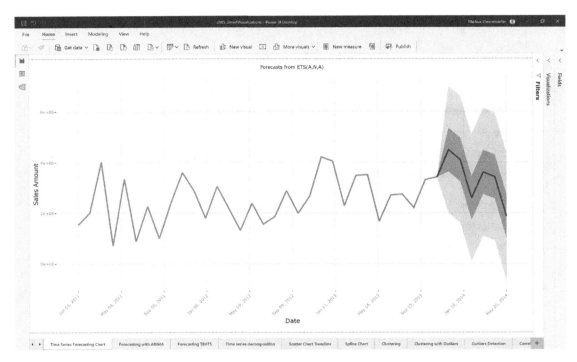

Figure 5-12. *Time series forecasting chart*

This chart offers you the flexibility to influence the following *Forecast* settings:

- *Forecasting settings*: *Forecast length, Trend component (Automatic, Multiplicative, Additive, None), Trend with damping (Automatic, TRUE, FALSE), Remainder component (Automatic, Multiplicative, Additive), Seasonal component (Automatic, Multiplicative, Additive),* and *Target seasonal factor (Automatic, Hour, Day, Week, Month, Quarter, Year)*

- *Confidence intervals* can be disabled. If enabled, choose out of a list of values between 0 and 0.999 for two intervals.

- *Graphical parameters*: *History data color, Forecast data color, Opacity, Line width*

- *Additional parameters*: *Show info* (at top of chart) and *Font size*

- *Export data*: If toggled to *On*, a button on the upper left appears to allow the user to export not only the actual data, but the forecasted data as well.

I leave all parameters on their default values (*Automatic*), except for a *Forecast length* of 6, *Additive Seasonal component* and *Month* as a *Target seasonal factor*. I changed the first *Confidence Interval* to 0.5 (50 percent) and left the second one on the default of 0.95 (95 percent). In section *Graphical parameters* I made sure to select the same colors I have in the previous examples.

Forecasting with ARIMA

As you might guess from the name of this visual, it is an implementation of auto-regressive integrated moving average (ARIMA) modeling to do forecasting (Figure 5-13). It is based on the time series algorithms in R package *zoo*.

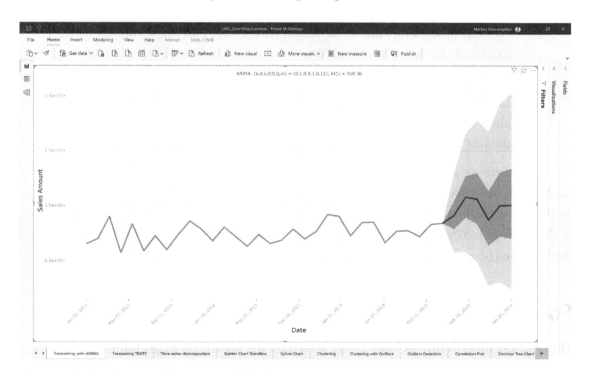

Figure 5-13. *Forecasting with ARIMA*

This visual allows for plenty of parameters with which to tune the forecast:

- *Forecasting settings*: *Forecast length*, two *Confidence levels* (between 0 and 0.999)

- *Seasonality*: *Target seasonal factor* (*Automatic, Manual, Hour, Day, Week, Month, Quarter,* or *Year*) and *Frequency*

- *Model Customization* (the parameters for the ARIMA algorithm): *Maximal p, Maximal d, Maximal q, Maximal P, Maximal D, Maximal Q, Allow drift, Allow mean, Box-Cox transformation* (*off, automatic,* or *manual*), and *Stepwise selection*

- *User defined model*: If enabled you can select values for *p, d, q, P, D,* and *Q* directly (and not only their max values).

- *Graphical parameters*: Choose *History data color, Forecast data color, Opacity,* and *Line width*

- *Info*: Shows the model parameters in the headline (according to the chosen *Font size* and *Text color*). You can enable to show goodness-criteria Akaike Information Criterion (*AIC*), Bayesian Information Criterion (*BIC*), or Corrected Akaike Information Criterion (*AICc*) for the model as well.

- *Export data*: If toggled to *On*, a button on the upper left appears to allow the user to export not only the actual data, but the forecasted data as well.

To achieve what I think is a useful forecast for the volatile reseller sales amount, I enabled *Seasonality*, chose *manual* as *Target seasonal factor*, set *Frequency* to 12, and enabled *User defined model* and set all criteria to 0, except for *d* and *D*, which I set on 1. In the settings in *Graphical parameters* I made sure that the chart had the same colors as the previous ones.

Forecasting TBATS

This visualization performs forecasting, based on the algorithm Trigonometric, Box-Cox Transform, ARMA Errors, Trend, Seasonal (TBATS) available in R packages *forecast* and *zoo*. For the demo data, this algorithm performed worse. Despite using several combinations of parameters, it did not discover the seasonality of the data and showed a flat line as the forecast, as you can see yourself in Figure 5-14. I kept this example to show you that not all algorithms will always provide useful forecasts for the available set of data under all circumstances. Always try different algorithms and hyper-tune them (a.k.a. set different parameters) before you derive any conclusions.

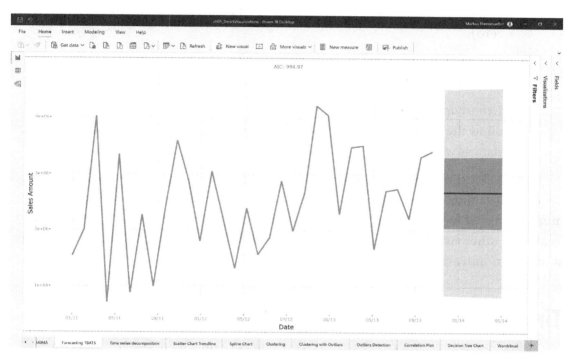

Figure 5-14.   *Forecasting with TBATS*

These are the parameters you can set:

- *Forecasting settings: Forecast length, Seasonal factor #1 (None, Manual, Hour, Day, Week, Month, Quarter, Year), Seasonal factor #2*

- *Confidence intervals* (two levels) between 0 and 0.999

- *Graphical parameters: Opacity, Line width, Actual data color, Forecast data color,* and *Show subset of dates (All, Last hour, Last day, Last week, Last month, Last year). Show fitted values* shows the forecasted values for the past as well. If you turn this on in the example file, you will get a relatively flat curve, confirming that the model totally ignores seasonality.

- *Labels and axes*: Labels color, Labels font size, Dates format on X-axis (*auto; 2001; 12/01; Jan 01, 2010; 01/20/10; 20/01/10, Jan 01; 01/20/10 12:00; Jan 01 12:00; 12:00; 2010,Q1; Thu Jan 20*), Scientific view Y-axis.

- *Advanced parameters* to turn *Positive data values* on or off. Turning this to *On* avoids a prediction of negative values (which in the case of sales amount makes sense, as the lowest possible sales amount is zero).

- *Info parameters*: Information content (*None, AIC, Cumulative, Method*) and its appearance (*Fonts size, Font color, Number of digits*)

- *Export data*: If toggled to *On*, a button on the upper left appears to allow the user to export not only the actual data, but the forecasted data as well.

I failed in setting parameters that let the algorithm discover the seasonality of the data. Disabling *Scientific view Y-axis* did not change anything in my example (the numbers stayed formatted in scientific formatting). I changed the appearance though, to match the other visuals. And I changed *Dates format on X-axis* from *auto* to *12/01* to only show the month and year.

Time Series Decomposition Chart

In case you are wondering how a time series prediction (like earlier) is done, this chart is for you. It is based on R packages *proto* and *zoo* and splits up the different parts of the forecast into the following, as you can see in Figure 5-15:

- *Data*: the actual values

- *Seasonal*: discovered seasonality (it shows a repeating pattern)

- *Trend*: over the whole time period

- *Remainder*: Difference of forecast (trend and seasonality) to actual value; that's the part not explained by the model; the better the model, the lower the remainder

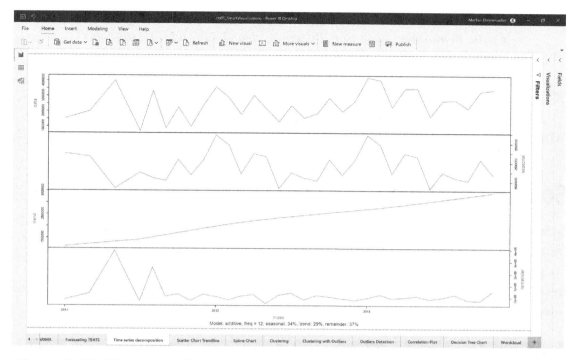

Figure 5-15. *Time series decomposition chart*

The information at the bottom of the chart explains that the model is an additive one (that was automatically discovered) and the seasonal frequency is 12 (which I manually set). The model explains 34 percent as seasonal value, 29 percent as a trend. Thirty-seven percent can't be explained by either (remainder). A big percentage of the remainder is the result of an "unseasonal" form of the actual data during year 2011. Of course, we do not know if 2011 was atypical or if the time series algorithm emphasized the years 2012 and 2013 too much.

You can change the calculation and appearance of this chart via the following *Format* options:

- *Time series model*: *Decomposition model (Additive, Multiplicative, Automatic), Seasonal factor (Autodetect from date, None, Manual, Hour, Day, Week, Month, Quarter,* or *Year)*

- *Algorithm parameters: Degree, Robust to outliers, Trend smoothness*

- *Graphical parameters*: *Plot type (Decomposition, Trend, Seasonal, Clean, Remainder, By season, By season clean), Line width, Line color, Labels color,* and *Labels font size*

- *Show information*: *On* or *Off* and *Font size* and *Text color*

To achieve the chart as shown, I adopted the following options: In *Time series model* I changed *Seasonal factor* to *Manual* and set *Frequency* to 12.

Scatter Chart with Trendline

Now we are back to a standard visual (scatter chart; Figure 5-16), to which I added a trendline in the *Analytics* pane (identical to what you saw for a line chart earlier). The chart shows the order quantity (on the y-axis) per unit price (on the x-axis). The dashed trendline shows an expected correlation: products with a higher unit price are ordered in lower quantities.

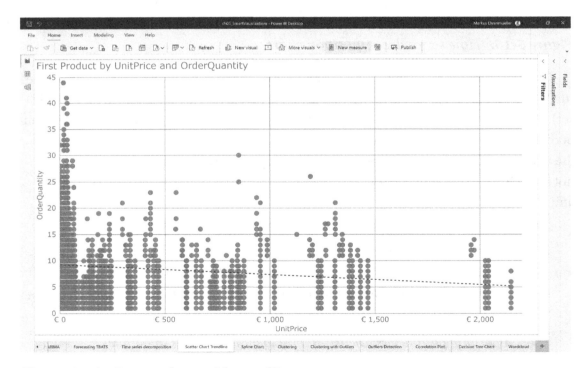

Figure 5-16. *Scatter chart with trendline*

Spline Chart

The trendline in a standard chart is calculated as a simple regression line, as we have already learned in the section about line charts. A trendline in the form of a straight line only gives us a rough direction. If we expect more of a forecast, the line has to be smoothed onto the actual values. The spline chart in Figure 5-17 gives us exactly that, with a little help from R package *graphics*.

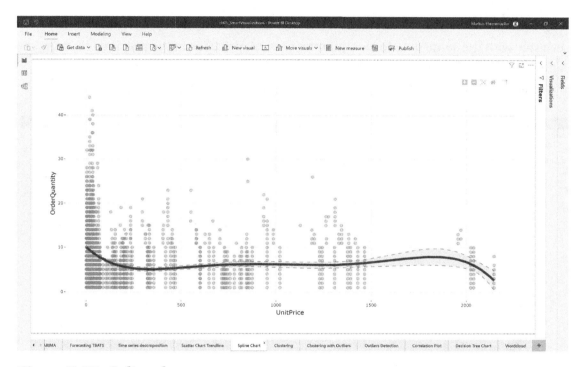

Figure 5-17. *Spline chart*

This chart allows us to set specific parameters, as follows:

- *Spline settings*: Besides the *Line color* we can choose the *Model* (*auto*, local regression [*loess*], generalized additive model [*gam*], polynomial regression of different degrees (*lm_poly1* to *lm_poly4*]) and the *Smoothness* of the curve

- *Confidence level* can be enabled or disabled and set to a free value.

- *Scatter plot* allows you to change the *Point color, Point size,* and *Opacity* of the points. *Sparsify* is turned on by default and helps you to visually identify dense areas.

I set *Model* to *lm_poly5* and changed the default colors to match the colors in the other charts we have created so far.

Clustering

Let's stay with the same data (order quantity per unit price) as in the previous section. Figure 5-18 does not show a straight or smooth trendline but instead visualizes clusters in the data discovered by applying the functionalities of several R packages.

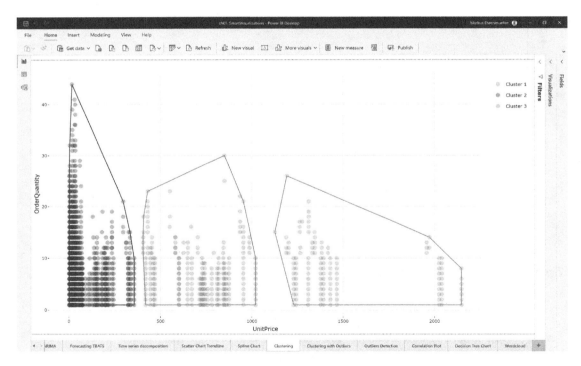

Figure 5-18. *Clustering*

The legend shows generic names for the clusters (as the algorithm lacks domain knowledge to find a correct name for the automatically discovered clusters). If you click on the cluster name once, you can toggle whether its data points are hidden or shown. If you double-click on the name, the data points for all other clusters are hidden or shown again.

The Clustering visual comes with plenty of options, as follows:

- *Data preprocessing* offers to *Scale* data (to standardize the value range of a column) and to *Apply PCA* (Principal Component Analysis), useful for high-dimensional data.

- *Clusters definition*: The *Number of clusters* can be set to *Auto* or a number between two and twelve. *Method* can be *Moderate, Slow,* or *Fast.*

- *Visual appearance* lets you choose the *Point opacity, Point size,* and if you want to *Draw ellipse, Draw convex hull,* or *Draw centroid.*

- *Points labeling* and *Cluster representative labeling* allow you to add and format labels at the data points.

- *Legend* can be enabled/disabled, and a color *Palette type* can be chosen.

- *Advanced*: Enter a value for *Minimum clusters, Maximum clusters, Maximum iterations,* and *Number of initializations. Sparsify* is turned on by default and helps you to visually identify dense areas.

I only enabled *Draw convex hull* and left all other options at their default values.

Clustering with Outliers

This visual is very similar to *Clustering,* described in the previous section. It applies a k-means algorithm onto the data (with the help of R packages *fpc* and *dbscan*). A significant difference is that it does not assign every single data point into a cluster, but rather identifies some of them as outliers. In Figure 5-19, the outliers are represented as grey checkmarks.

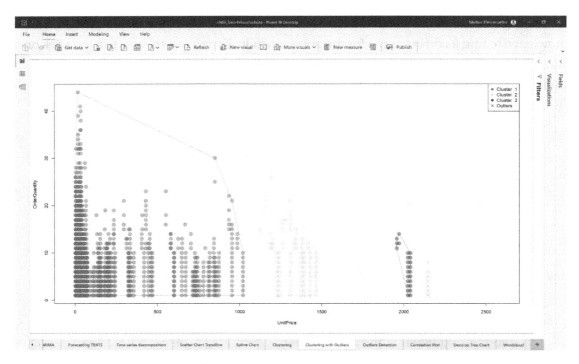

Figure 5-19. *Clustering with outliers*

Most of the options are identical to the Clustering visual, as follows:

- *Data preprocessing* offers to *Scale* data (to standardize the value range of a column) and to *Apply PCA* (Principal Component Analysis), useful for high-dimensional data.

- *Clusters definition*: The *Granularity* method can be set to *Auto, Manual,* or *Scale. Manual* lets you set a number for epsilon (*eps*). The smaller the number, the fewer clusters (and outliers) you will get. When you select *Scale* you can choose a percentage between 0 and 100. The higher the percentage, the fewer clusters there will be. *Find minimum points automatically* can be enabled or disabled. If disabled you can define on your own how many data points are needed at minimum to form a cluster. Clusters below the minimum are considered outliers. The lower the number, the more outliers you will get.

- *Visual appearance* lets you choose the *Point opacity, Point size,* and if you want to *Draw ellipse, Draw convex hull,* or *Draw centroid.*

- *Points labeling* and *Cluster representative labeling* allow you to add and format labels at the data points.

- *Legend* can be enabled/disabled, and a color *Palette type* can be chosen. The list of palettes is smaller than the one for the Clustering visual.

- *Additional parameters* allow you to enable or disable to *Show warning.*

I did set *Granularity method* to *Manual* and epsilon to 100 and turned *Find minimum points automatically* off and set the *Minimum points per seed* to 8. This will find four clusters in my data and will mark data points beyond a unit price of 2100 as outliers. I enabled *Draw convex hull* in *Visual appearance* to get a border around the clusters. The algorithms in Clustering and Clustering with Outliers seem to be a bit different, as I did get different clusters in each.

Outliers Detection

Outliers Detection is like the previous chart, minus clustering: It concentrates solely on detecting outliers, calculated with R packages *DMwR* and *ggplot2.* Figure 5-20 shows us outliers, which mostly have higher order quantity, except for some high-priced items where low order quantity is an outlier as well.

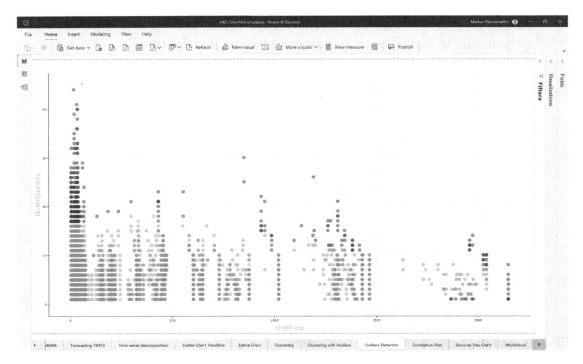

Figure 5-20. *Outliers Detection*

This visual has many possibilities to tweak the algorithm and the visual appearance of the chart, as follows:

- *Detection* allows you to set *Outlier type* (to *Peaks and subpeaks, Subpeaks,* or *Peaks*), *Algorithm* (*Manual, Zscore, Tukey, LOF or Cook's distance*), and to enable/disable *Scale (if applicable).*

- *Visualization*: Chose either a *Scatter, Boxplot,* or *Density* as the *Plot type* and decide if you want to *Visualize outlier's score.*

- *Markers* let you change *Inliers color, Outliers color, Point size,* and *Opacity.*

- *Axes* let you change *Labels color, Labels size, Size ticks, Scale X format (None, Comma, Scientific, Dollar),* and *Scale Y format (None, Comma, Scientific, Dollar)*

To get the chart in Figure 5-20, I set the following options: In *Detection* I chose *Cook's distance* and set the *Threshold* to 2. I changed the colors in *Markers* to fit the usual colors. Obviously, the algorithm here is different than the one from Clustering with Outliers, as different data points are marked as outliers.

Correlation Plot

A correlation plot can show you the correlation of different measures, and optionally cluster them by correlation coefficient. The visual uses R package *corrplot*. In the next example (Figure 5-21), we look out for correlations between different measures of reseller sales: discount amount, freight, order quantity, product standard cost, sales amount, tax amount, total product cost, and unit price average.

Note   I added column *OrderDate* to the values of the correlation plot, even if a date column is not plotted. The only reason I added this column is to make sure that Power BI calculates aggregates per *OrderDate* before the visual calculates the correlation. If I had omitted *OrderDate*, Power BI would aggregate all measures to a single total line. From only one (aggregated) row it is hard to reveal the true correlation. It might make a big difference depending on which additional column you add to the values for the visual. If you put *SalesOrderNumber* into the list of values, you will not see any negative correlation with unit price average, as this correlation is only visible on *OrderDate*'s granularity.

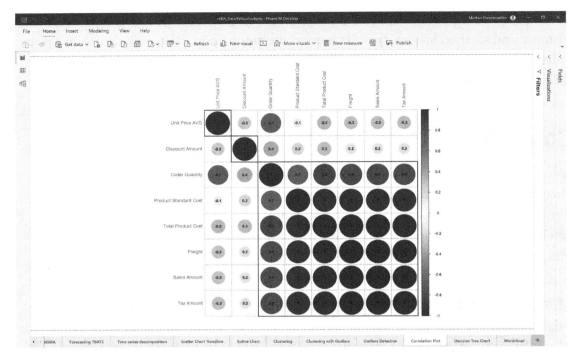

Figure 5-21. *Correlation plot*

A correlation of +1 between two measures means that if one measure increases by one, the other increases by one as well (and the other way around). A correlation of -1 between two measures means that if one measure increases by one, the other decreases by one (and the other way around). A correlation of 0 means that if one measure increases by one, the other does only react randomly to that. The other measure might increase, decrease, or not change at all.

Figure 5-21 shows blue and red dots in a raster with all the preceding mentioned measures in rows and columns. A blue dot means a positive correlation. A red dot means a negative correlation. The size and hue of the dot indicate how strong the correlation is. A big dark blue dot is a sign for a strong positive correlation. As all measures have a correlation of +1 with themselves, we get big dark blue dots in a diagonal from top left to bottom right.

The plot reveals that there is weak negative correlation for unit price average and all the other measures, except for order quantity, where there is a strong negative correlation. We see a weak positive correlation between discount amount and all other measures, except for unit price average, with which there is a weak negative correlation. The other measures have a rather strong positive correlation with each other.

You cannot change the algorithm behind the correlation plot, only change its appearance, as follows:

- *Correlation plot parameters*: Can be turned off to stick with the default values. If you enable this section, you can change *Element shape* (*Circle, Square, Ellipse, Number, Shade, Color,* or *Pie*), and let it *Draw clusters*.

- *Labels*: Here you can change *Fonts size* and *Color*.

- *Correlation coefficients*: Enable this option to display a value between -1 and +1, representing the correlation coefficient. Change # *digits, Color,* and *Font size*.

- *Additional settings* allow you to *Show warnings*.

I did largely stick with the default options, except for these ones: I enabled *Correlation plot parameters* to set *Draw clusters* on *Auto* and enabled *Correlation coefficients*.

Decision Tree Chart

A decision tree is a machine learning model that analyzes correlations in the data and can be visualized as a tree. The Decision Tree Chart custom visual is based on R package *rpart* to build the model and *rpart.plot* to visualize the model (tree-like). Like many trees in IT or mathematics, this tree is drawn upside-down, with the trunk on top and the leaves at the bottom. The most interesting part of the tree is the leaf-level, which is on the bottom of the chart. I visualized a simple decision tree (with only four leaves and two intermediate levels) in Figure 5-22.

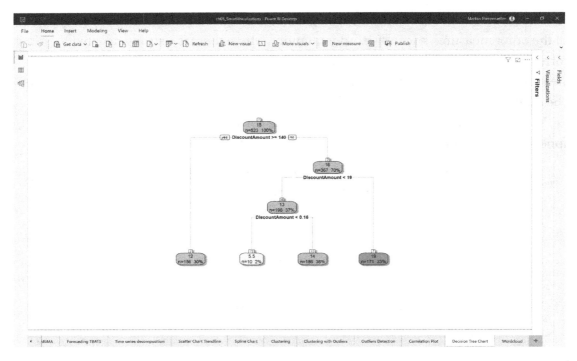

Figure 5-22. *Decision tree*

This decision tree predicts the order quantity as a *Target Variable* dependent on the *Input Variable* discount amount. Each node (trunk, branch, or leaf) shows four numbers. Let's concentrate on the bottom left one. On top, outside of the green rectangle, there is the node number in tree (2, in our case); not all nodes are shown in the visualization. In the first line inside the green rectangle, we get the predicted value for the target variable (order quantity): 12. In the second line inside the rectangle, we see the number of rows that are predicted with this value: 156 rows, which is 30 percent of all the rows available in the data set.

For the outcome of a prediction with a decision tree, only the leaf-level nodes (plotted on the bottom) are used. Nonetheless, predicted values for the trunk and branches are shown as well. From them you can learn what the prediction would be with just part of the information considered. Look, for example, on the trunk. No decision on any *DiscountAmount* is done at this point. Then the prediction would be an order quantity of 15. The more information is used, the more specific prediction will be done (16 or 13 on branch nodes 3 and 6).

You might set the following options to change the algorithm:

- *Tree parameters*: *Maximum depth* (a value between 2 and 15 to limit the amount of levels from trunk to leaf) and *Minimum bucket size* (a value between 2 and 100; the higher this number, the lower the number of nodes)

- *Advanced parameters*: *Complexity* (a number between 0.5 and one trillion to control if node should be further split or not), *Cross-validation* (*Auto, None, 2-fold* to *100-fold*; the higher the value the better the accuracy, but the longer the calculation), and *Maximum attempts* (a number between 1 and 1000; the higher the number, the better the accuracy, but the longer the calculation)

- *Additional parameters* to toggle *Show warning* and *Show info*.

I didn't change any of the default values for this example.

Word Cloud

I always wondered what kind of insights I could gain from the English description in *Adventure Work*'s product table. This column contains a distinct description for every single product. Listing the descriptions and reading through them to learn about the product is not a very satisfying task.

In Figure 5-23, I built a word cloud on this column. It perfectly sums the content of the column up. The more often a word is mentioned, the more prominently it is shown in the visualization through a bigger font. Obviously, over all products, terms like "frame," "aluminum," "bike," or "lightweight" are very common in the descriptions.

Figure 5-23. *Word cloud*

Here are the options of this custom visual:

- Beside the *Category* column, for which you will show the content, you can add a column for *Values* in the *Fields* pane. The word cloud is then built upon the weight of the *Value* instead of on frequency.

- *General*: *Minimum number of repetitions to display, Max number of words, Min font size, Max font size, Word-breaking* (if turned off, the whole text is displayed instead of single words), and *Special characters*.

- *Stop words* allows you to provide a custom list of words you do not want to see in the word cloud.

- *Rotate Text* can be enabled or disabled. If enabled, you can set the *Min Angle, Max Angle,* and *Max Number of orientations.*

- *Performance* can be set to *Pre-estimate words count to draw* and its *Quality.*

I added my own list of *Default Stop Words* and turned *Rotate Text* off.

We will welcome back the word cloud visual in later chapters (about R and Azure's services).

Key Takeaways

You have learned a lot in this chapter:

- Standard visuals offer to add a trendline or a forecast in the *Analytics* pane.

- By writing your own formula in DAX, you can calculate a simple regression on your own. This gives you full control over the calculation formula and filters.

- You can load custom visuals into your Power BI Desktop file. These offer functionalities beyond the standard visuals, without the need to write code. We only looked at a small number of custom visuals.

- We forecasted values over time with the help of custom visuals Time Series Forecasting Chart, Forecasting with ARIMA, Forecasting TBATS, and Time Series Decomposition Chart.

- We learned about trends in a dataset with Spline Chart.

- We automatically clustered a dataset with the help of Clustering, Clustering with Outliers, and Outliers Detections.

- We saw further examples with Correlation Plot, Decision Tree Chart, and Word Cloud.

## CHAPTER 6

Experimenting with Scenarios

Have you ever had the challenge of not just showing the data as it is, but of also going over different possible scenarios, like the worst, the best, and the most likely cases? In Power BI, you can easily play around with values, which allows you to do exactly that: change the parameter and see how this will change dependent measures. We will use this feature to find the best price for your products in order to maximize your sales amount.

Scenarios in Action

I have prepared an example where we can play with different discounts on our prices to enable us to find the optimum price that increases sales to the possible maximum. The model (to be precise: DAX measures) behind the example applies the concept of *price elasticity on demand*, which assumes the following:

- The lower the prices, the higher the quantity bought by our customers. The higher the prices, the lower the demand by our customers.

- How much the quantity increases on a given decrease of the price depends on the product's elasticity. An elasticity of 1 means that a 10 percent decrease in price will increase the quantity sold equally (by 10 percent). An elasticity of 2 means that a 10 percent decrease in price will increase the demand by 20 percent.

- We have control over the price, while the elasticity for a certain product is given by the market.

© Markus Ehrenmueller-Jensen 2020
M. Ehrenmueller-Jensen, *Self-Service AI with Power BI Desktop*, https://doi.org/10.1007/978-1-4842-6231-3_6

This concept has limitations. The quantity will not increase endlessly (e.g., even with a very low price, you will not start drinking several gallons of milk a day), just to mention one. I took elasticity only as an example to demonstrate how you can build parameterized reports in which calculations react dynamically to changes of user selection.

This is what the sample report (Figure 6-1) offers:

- You can select or input a percentage value for a *Discount* on the prices.

- You can select or input a value for the price *Elasticity* of demand on the products.

- The report shows the current value of *Sales Amount.*

- The report shows the calculated new values (after the *Discount* applied): *New Order Quantity, New Unit Price, New Sales Amount.*

- The report shows the difference of the current to the new sales amount (*Sales Amount Delta*). A positive value means that the price change resulted in a higher sales amount.

- The line chart in Figure 6-1 shows the difference of the current to the new sales amount (*Sales Amount Delta*) on the y-axis over possible values of *Discount* (between 0 percent and 100 percent) on the x-axis.

- A dashed line marks the highest possible difference in sales amount for the given elasticity. The maximum difference in sales amount is shown separately (*Sales Amount Delta Maximum*) on the top right of the report page.

- Also included is the *Best Discount*, which will lead to the maximum difference in sales amount (assuming that the elasticity and the model are correct).

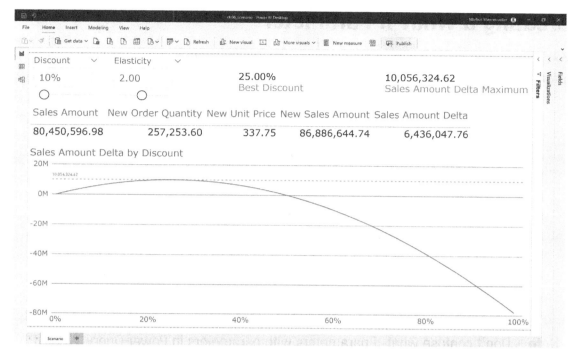

Figure 6-1. *Example report showing the capabilities of two what-if parameters*

If you change the *Discount*, you will see that both the *New Order Quantity* and *New Unit Price* will change as well. Increasing the *Discount* will increase the quantity but decrease the price. This will further lead to a different *New Sales Amount* and a change in the *Sales Amount Delta*. The rest of the report stays the same.

If you change the *Elasticity*, you will see that only the *New Order Quantity* will change. Increasing the *Elasticity* will increase the quantity. This will further lead to a different *New Sales Amount* and a change in the *Sales Amount Delta*. The rest of the report stays the same.

In the following sections, I will guide you step-by-step to rebuild this example.

Creating a What-if Parameter

You create a new what-if parameter via *Modeling* ➤ *New parameter* and provide some information:

- *Name* of the what-if parameter

- *Data type* (*Whole number, Decimal number, Fixed decimal number*)

- *Minimum*

- *Maximum*

- *Increment*

- *Default*: a value used for the what-if parameter when no selection in the slicer is yet made

- *Add slicer to this page*

Note   Don't confuse what-if parameters with parameters in Power Query. What-if parameters are only available in the report and can be used to change the behavior of DAX calculations. Power Query parameters are currently not available inside a report (but only if you open Power Query). A Power Query parameter changes the behavior of a query.

For this example, I created two what-if parameters:

- One with name *Discount*. This will allow you to choose a percentage, and therefore I chose *Decimal number* as the *Data type* and a *Minimum* of 0 and a *Maximum* of 1. The *Increment* is 0.01 (to allow one to change the what-if parameter in single percentage steps). I want the report user to explicitly select a value; therefore, I didn't enter anything in *Default*. I let the wizard *Add a slicer to this page* (Figure 6-2).

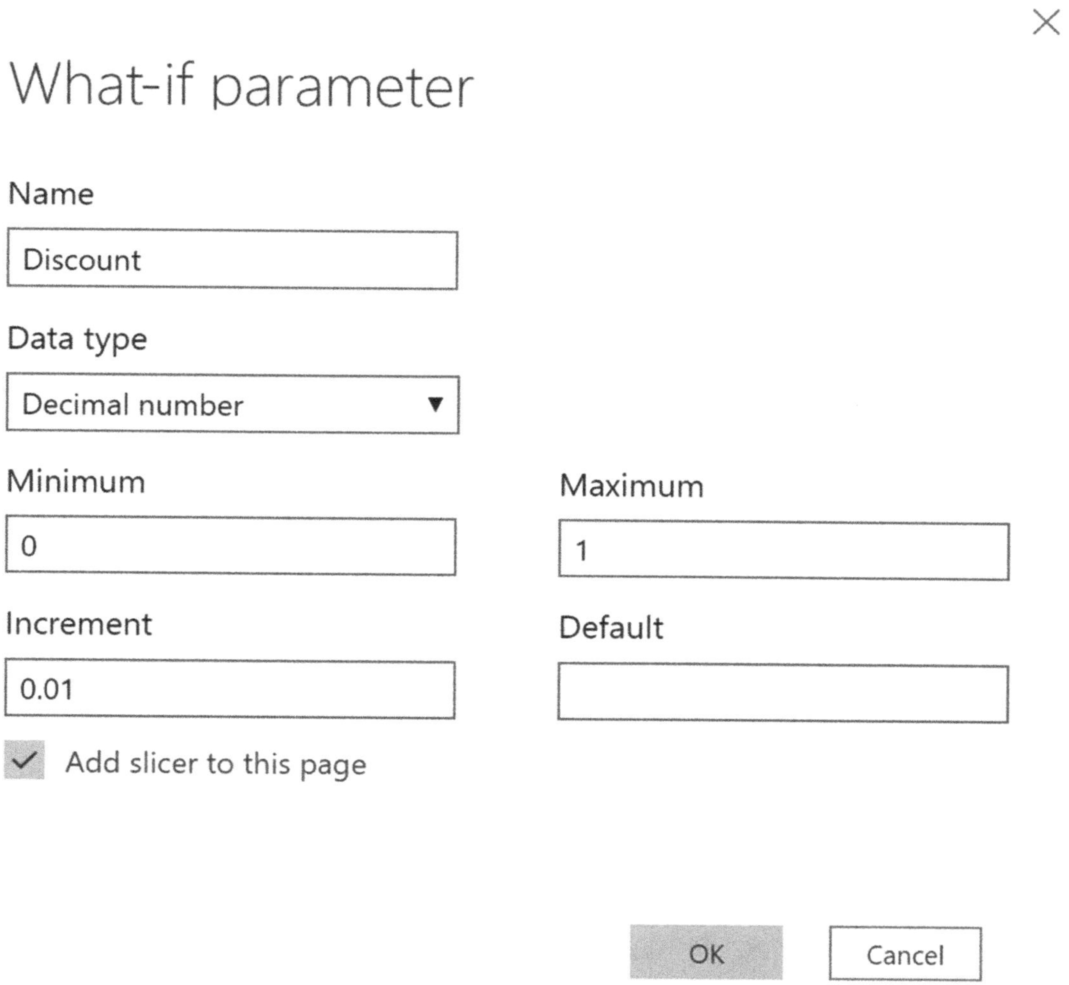

✕

What-if parameter

Name

Discount

Data type

Decimal number   ▼

Minimum

0

Maximum

1

Increment

0.01

Default

✓ Add slicer to this page

OK   Cancel

Figure 6-2. *Definition of what-if parameter Discount*

- A second one with name *Elasticity*. This will allow you to choose the *price elasticity of demands* overall on our products. I chose *Decimal number* as the *Data type* and a *Minimum* of 0 and a *Maximum* of 5. The *Increment* is 0.1 (to allow one to choose to be very granular). I want the report user to explicitly select a value; therefore, didn't I enter anything in *Default*. I let the wizard *Add a slicer to this page* (Figure 6-3).

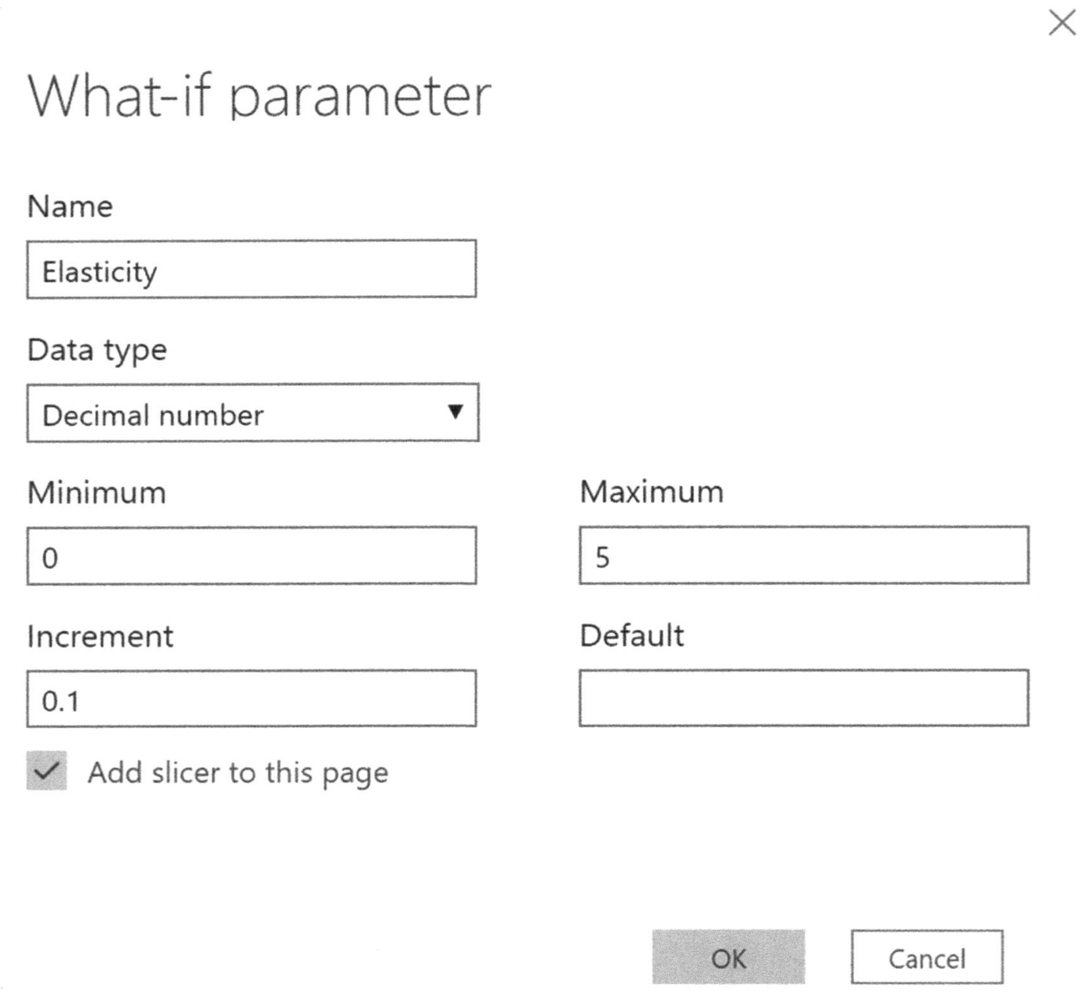

✕

What-if parameter

Name

| Elasticity |

Data type

| Decimal number   ▼ |

Minimum

| 0 |

Maximum

| 5 |

Increment

| 0.1 |

Default

| |

☑ Add slicer to this page

| OK | | Cancel |

Figure 6-3. *Definition of what-if parameter Elasticity*

The wizard then creates three things at once for you (per what-if parameter):

- A table with a range of values (with the table name and column name
  equal to the name of the what-if parameter), generated via DAX function
  GENERATESERIES, with the *Minimum, Maximum,* and *Increment* as its
  parameters. If you select the table name in the *Fields* pane on the very
  right you can see the DAX statement: Discount = GENERATESERIES
  (0, 1, 0.01) and Elasticity = GENERATESERIES(0, 5, 0.1). The
  result for *Discount* (after I changed the *Format* to *Percentage* for column
  Discount in the *Column tools* in the ribbon) is shown in Figure 6-4.

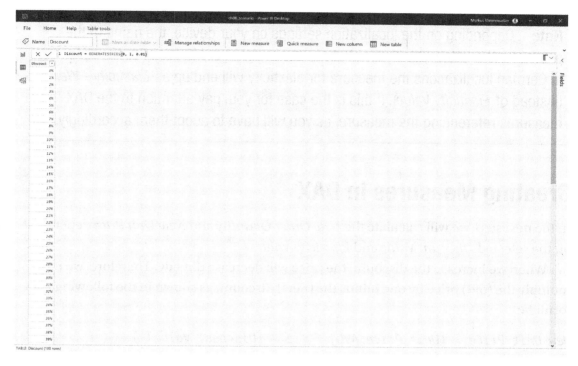

Figure 6-4. *Table Discount consists of 100 rows, representing the possible percentage values*

- A measure showing the selected value from the table (with the name of the what-if parameter with postfix "Value" added), generated via DAX function SELECTEDVALUE and the column from the table as a parameter with the optional *Default* value as a second parameter (if applicable): Discount Value = SELECTEDVALUE('Discount' [Discount]) and Elasticity Value = SELECTEDVALUE (' Elasticity '[Elasticity])

- A slicer on the report to make selections from the table (optional), shown in Figure 6-1

If you forgot to enable *Add slider to this page* or the wizard did not create a slicer for you, don't panic. Manually create a slicer instead. Click on an empty space on your report pane, choose the Slicer visual from the *Visualization* pane, and add the column (not the measure) from the automatically created table. In case of elasticity, you would add column *Elasticity* from the table *Elasticity* (and not the measure named *Elasticity Value*).

> **Note**   Depending on the localization settings on your device, the name of the
> postfix for the automatically created measure might be different. For example,
> in German localizations the measure for elasticity will end up as *Elasticity - Wert*
> (instead of *Elasticity Value*). If this is the case for you, pay attention to the DAX
> measures referencing the measure, as you will have to adopt them accordingly.

Creating Measures in DAX

In the next step, we will calculate the *New Order Quantity* and *New Unit Price* dependent
on the selected *Discount Value* and *Elasticity Value*.

When we increase the discount, the price will decrease directly. Therefore, we must
multiply the (old) price by one minus the (new) discount, as shown in the following DAX
formula:

New Unit Price = [Unit Price AVG] * (1 - [Discount Value])

The calculation for the quantity is slightly more complex. The (new) quantity is
not influenced by the *Discount Value* only, but by the *Elasticity Value* as well. When we
increase the discount, the quantity will increase as well. The increase is factored by the
elasticity. I multiply the (old) quantity by 1 plus the (new) discount multiplied by the
elasticity:

New Order Quantity = [Order Quantity] * (1 + [Discount Value] *
[Elasticity Value])

Sales amount is on the simple side again: (new) quantity times (new) price equals
(new) sales amount:

New Sales Amount = [New Order Quantity] * [New Unit Price]

And finally, we calculate the difference between the (current) sales amount and the
new one, to see if it makes sense to lower the prices:

Sales Amount Delta = [New Sales Amount] - [Sales Amount]

I added the (current) sales amount and all four new measures in a multi-row card
beneath the slicers (see Figure 6-1).

Ahead of the Curve

Next, I built a line chart with *Sales Amount Delta* on the *Values* and *Discount* on *Axis*. As the *Discount*'s slicer is filtering all visuals on the report page, it filters the line chart as well, and we end up with only a single data point at exactly the selected discount in Figure 6-5. If you change the *d*iscount, the data point in the chart will move up or down.

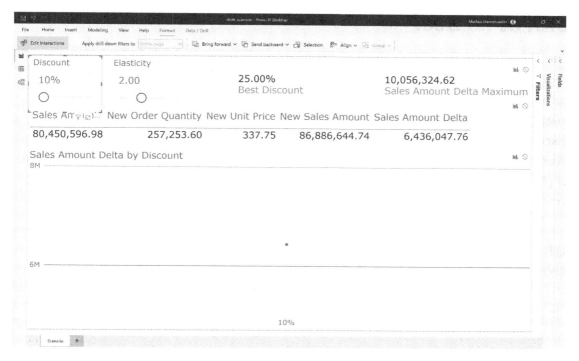

Figure 6-5. *A line chart for a single data point*

To get a line chart over the whole range of discounts (from 0 percent to 100 percent), we can change the slicer to not filter the line chart. Click first on the *Discount* slicer and then select *Format* ➤ *Edit interactions* in the ribbon (shown in Figure 6-5). All other visuals now show two icons on their upper right corner: the icon with column charts and funnel is pre-selected. Click on the circle with the crossing line; a curved line showing the *Sales Amount Delta* measure over all percentage values appears now in the line chart (as in Figure 6-1).

In the *Analytics* options of the line chart you can add a *Max line* to point out what the maximum sales amount difference could be at the given elasticity. You can turn *Data label* on, but unfortunately we cannot change the font size—the number on the left at the beginning of the dashed maximum line is hard to read. Fortunately, with a bit of DAX magic, we can calculate the number on our own. I'll show you in the next section.

DAX at Its Best

The following DAX measure gives us the maximum [Sales Amount Delta] calculated over all discounts. The ALL function wrapped around 'Discount'[Discount] makes sure that any existing filter on column 'Discount'[Discount] outside the measure (e.g., because of a selection in the slicer for *Discount*) is ignored and all possible discount values are taken into account. MAXX (no, the double X is not a typo) iterates over all rows returned from ALL ('Discount'[Discount]) and returns the maximum of [Sales Amount Delta] (yes, there is also a MINX, SUMX, AVERAGEX, or PRODUCTX function available).

That's the measure:

```
Sales Amount Delta Maximum =
MAXX (
    ALL ( 'Discount'[Discount] ),
    [Sales Amount Delta]
)
```

I put this measure in a multi-row card and put it on the top right of the report page. In my example, I accompanied the maximum sales amount with the *Best Discount*. That's the discount percentage at which we achieve the [Sales Amount Delta Maximum]. First, I save the result of [Sales Amount Delta Maximum] in variable SalesAmountDeltaMax. The first parameter for MINX is a call to function FILTER. Its first parameter is again ALL ('Discount'[Discount]), a list of all possible discounts. This list is filtered on only those rows where the [Sales Amount Delta] equals the value of variable SalesAmountDeltaMax. The second parameter for MINX is 'Discount'[Discount]. The sole purpose for having MINX in the code is to make sure that the code returns only a single discount value in case several discounts end up in the maximum sales amount delta (and this single value will be the lower one).

Here is the full measure:

```
Best Discount =
VAR SalesAmountDeltaMax = [Sales Amount Delta Maximum]
VAR DiscountsWithMaxSales =
    MINX (
        FILTER (
            ALL ( 'Discount'[Discount] ),
            [Sales Amount Delta] = SalesAmountDeltaMax
        ),
        'Discount'[Discount]
    )
RETURN
    DiscountsWithMaxSales
```

Adding this measure to the multi-row card on the top right of the report finishes this example.

Key Takeaways

The most important stuff from this chapter is as follows:

- Price elasticity of demand is an economical concept to explain how the demanded quantity changes dependent on the price of the offered products and services. The concrete price elasticity of a product is determined by the consumers and will be different per product (group).

- What-if parameter is a wizard that creates a table, a measure, and a slicer at once to enable the report user to change a numeric value, which can be part of DAX calculations. Depending on the DAX calculation, the use cases for what-if parameters are plenty. You can inject a what-if parameter into an R or Python visual (see Chapter 9) as well.

- DAX is the language in Power BI to create calculated columns, measures, and calculated tables. A what-if parameter creates a measure that contains the selected value. We can write further calculations based on the what-if parameter to show the report user the consequence of the changed value.

- *Edit interactions* is the way to disable (and re-enable) the cross-filter functionality. Cross-filter means that a selection in one visual (or slicer) filters the content of the other visuals on the same report page. In cases where this functionality is not expected by the report user or is unwanted by the report creator (like in our line chart), we can turn this off.

CHAPTER 7

Characterizing a Dataset

When I get handed a new set of data, I want to get a quick overview of what kind of data to expect before I start building the data model and reports. With just a few clicks you will get descriptive statistics for the dataset: for example, the count, the minimum and maximum of the available values, average, and standard deviation. You can easily visualize the value distribution and the amount of missing values and gain insights about the data even before you build your first report. Part of this metadata can be loaded into the data model, and you can build reports on it, if you want.

Power Query

In this chapter, we concentrate on a tool inside Power BI Desktop: Power Query. Select *Home* in the ribbon and then *Transform Data* (in the *Queries* section) to open Power Query. Power Query is part of Power BI Desktop, but shows up in a separate window, which helps if you have a big enough monitor (to work in Power Query and in Power BI Desktop simultaneously) but can be confusing in the beginning.

Note A typical difficulty for beginners in Power BI is to distinguish Power Query from the Power BI Data view, as the centers of the screens resemble each other. Some people turn the font in Power Query to monospaced (in the *View* ribbon and the *Data Preview* section) to make the Power Query window and the Power BI Desktop Data view more easily distinguishable.

In Power Query, you maintain your connections to the data sources; filter, shape and transform the data; and have tools at hand to inform you about the quality of the data. Every query in Power Query ends up as a table in the data model in Power BI with the same name.

© Markus Ehrenmueller-Jensen 2020
M. Ehrenmueller-Jensen, *Self-Service AI with Power BI Desktop*, https://doi.org/10.1007/978-1-4842-6231-3_7

In Figure 7-1 I have opened Power Query. On the left side of the screen you get a (grouped) list of queries (e.g., query *Employee* in group *AdventureWorksDW*). On the right-hand side every one of the *Applied Steps* for the query are listed (e.g., *Source, Navigation*, and *Removed Columns* for query *Employee*). In the center, you see the query result: its columns (e.g., *EmployeeKey, ParentEmployeeKey, EmployeeNationalIDAlternat eKey*, …) and rows.

Power Query's graphical interface lets you fulfill even complex tasks to filter, shape, and transform the data with a just a few mouse clicks. In only rare cases will you need to maintain or even write your own Power Query scripts. We will write a very simple one in the section "Table Profile."

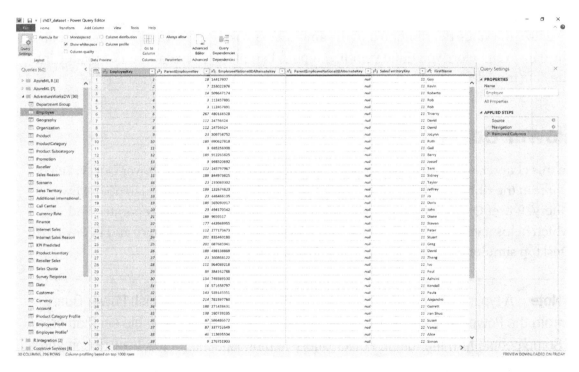

Figure 7-1. *Power Query*

Column Quality

In Power Query, select *View* in the ribbon and then make sure the checkmark at *Column quality* in the *Data Preview* section is ticked, as shown in Figure 7-2.

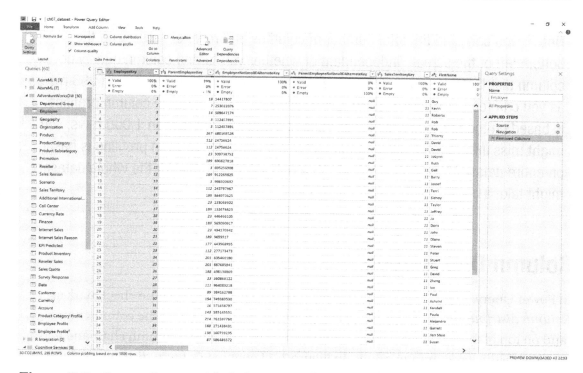

Figure 7-2. *Power Query with Column quality enabled*

Beneath the column name (and on top of the data rows) a new section is introduced with three lines showing a percentage (in relation to the total number of rows):

- *Valid* (green dot): percentage of rows not having an error or not empty

- *Error* (red dot): percentage of rows having an error

- *Empty* (grey dot): percentage of rows containing an empty string or *null*

The colors of the three dots are the same as the small bar/underline just underneath the column name. Because *EmployerKey* has 100 percent valid rows, the column name is underlined with a full-length green line. *ParentEmployeeNationalIDAlternateKey* contains empty rows only. It is underlined in grey. Column *ParentEmployeeKey* contains 1 percent empty and 99 percent valid rows. This is represented with an underline in mostly green and a small grey portion on the right end.

> **Note**   You can read the total number of columns and rows of a query on the
> bottom left of the screen (independent of whether you turned column quality or
> column distribution on or not). Query *Employee* contains 37 columns and 296 rows,
> as you can read in Figure 7-1. Here, it is also mentioned that the column profiling
> is based on the top 1,000 rows. For queries containing more than 1,000 rows, you
> might miss information. Click on the text to change it to "Column profiling is based
> on entire dataset." Be aware though, that building the profile for the whole dataset
> might take a considerable amount of time.

Column Distribution

In Power Query, select *View* in the ribbon and then make sure that the checkmark at
Column distribution in the *Data Preview* section is ticked. Beneath the column name
(and on top of the data rows) a new section is introduced with a column chart and a
count of distinct and unique values. In Figure 7-3 I enabled *Column distribution* (and
disabled *Column quality,* to make it clearer about what we are talking about in this
section).

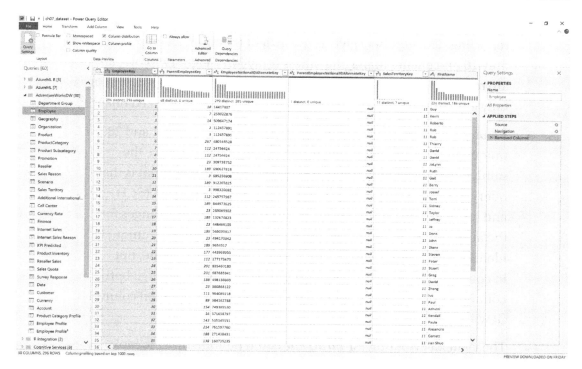

Figure 7-3. *Power Query with Column distribution enabled*

The count of *distinct* values gives you information on how many different values are found in this column. The count of *unique* values gives you information on how many rows have a value that only appears once in the whole query. If you move the mouse cursor over, a tooltip with both the absolute number and a percentage value is revealed. The percentage value gives us the relation to the overall count of rows in the table (which is shown on the bottom left of the screen).

We can gain the following insights from both counts:

- For *EmployeeKey*, both counts are the same (296) and they both cover 100 percent of the rows. That's expected, as this column is the primary key of the table and a primary key uniquely identifies every single row of the table. Different counts would mean that this column can't be a true primary key of the table.

- The numbers for *ParentEmployeeKey* are different: forty-eight distinct values (16 percent) and four unique (1 percent). This column contains the *EmployeeKey* of a different row in the same table to

161

point to the employee to whom the employee of the current row is reporting. The numbers give us the insight that from all employees only 16 percent are managing others (only forty-eight *EmployeeKey*s are referenced as *ParentEmployeeKey*). Four of the managers manage only a single employee (their *EmployeeKey* is only referenced once).

- *EmployeeNationalIDAlternateKey* shows only 290 distinct and 285 unique rows. That means that eleven rows (296–285) share the same *EmployeeNationalIDAlternateKey*. The reason for that is that query *Employee* is a so-called slowly changing dimension (SCD) and has *StartDate* and *EndDate* columns, which track changes for the attributes of an employee. All employee attributes are uniquely identified not with *EmployeeNationalIDAlternateKey* alone, but in combination with *StartDate*. To prove this, you can merge both columns and look at the column distribution of the merged column (it will be 296 distinct and 296 unique rows).

- The value distribution for *BirthDate* could be something to look closer at:

- Only 249 values (out of 296 rows) are unique. That means that a total of forty-seven (296 minus 249) have their birthday on the same date (not only the same day and month, but year as well; sharing the same day and month is more likely than you might think: https://en.wikipedia.org/wiki/Birthday_problem) as at least one other employee.

- The column contains 271 distinct (a.k.a. different) values. That means that twenty-two birthdates (271 minus 249) are shared by at least two employees. One birthdate is used 2.1 times (47 divided by 22) on average. Practically, this means that one or more birthdates are shared by more than two employees.

- In Figure 7-4 you see a Power BI report I generated to list and visualize the discussed numbers for column *BirthDate* to make the counts and deltas more understandable. Take a minute and make sure you understand each of the columns and their totals in both the table and the two column charts. For each existing value of column *BirthDate* we see the count of *Rows*, the number of *Distinct* rows, the number of *Duplicates* (if the value is not unique), the *Unique* count, and a flag if the value is *Not Unique* over the whole query. The *Totals* match the numbers just discussed and are shown in two column charts. One sums the *Distinct* and the non-distinct (a.k.a. *Duplicate*) *Rows* up, the other the *Distinct* and *Non-Unique Values*.

- The number of distinct and unique rows for *BirthDate* seem to be unlikely in a real-world scenario and would point out a possible data quality issue. In our case, it reveals that the demo data was generated without considering a realistic distribution in birth dates.

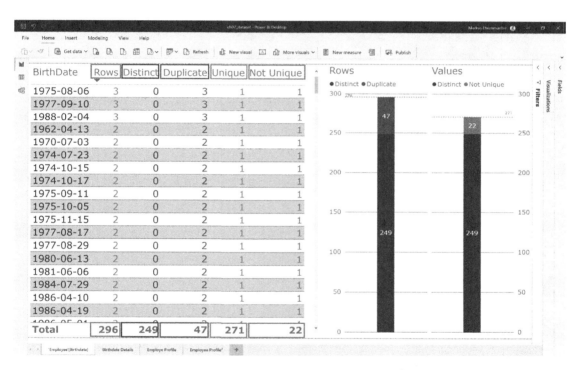

Figure 7-4. *Power BI report showing the different counts (and their deltas) for column BirthDate*

- With *LoginID* or *EmailAddress* we see the same problem as with *EmployeeNationalIDAlternateKey*. These columns are only unique in combination with *StartDate*.

- *MaritalStatus* looks good: We only allow for two different values (*M* = married and *S* = single)—and we have exactly two distinct values in this row. This is similarly true for columns *SalariedFlag*, *Gender*, *PayFrequency*, *CurrentFlag*, *SalesPersonFlag*, and *Status*.

163

- We have sixteen distinct (different) *DepartmentNames*. Every department has at least two employees assigned (0 unique values).

Quality and Distribution Peek

If you right-click the section containing the *Column quality* and/or *Column distribution* area, you can set the following actions to clean up your query (Figure 7-5):

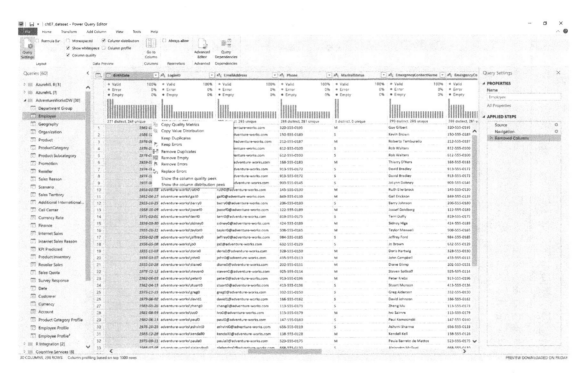

Figure 7-5. *Right-clicking the Column Quality/Column Distribution area discloses details and action items*

- *Copy Quality Metrics* (more details to come)

- *Copy Value Distribution* (more details to come)

- *Keep Duplicates*: Filters the whole query to only those rows that have values that are not unique in this column. In the case of *BirthDate* there are twenty-two rows.

- *Keep Errors*: Filters the whole query to only those rows that are having error values in this column. You would temporarily filter on the error rows to inspect them more closely, or permanently to build a report in Power BI revealing problems in the data. In the case of *BirthDate* this ends up with an empty query (as *BirthDate* has no errors).

To demonstrate the feature, I added a new column *Phone Area Code* in a new query *Employee with error*. I deliberately introduced an error into this column by first extracting the three first digits of the phone number and then changed the data type of the column to *Whole Number*. In four rows the phone numbers are prefixed with the international code "1" (instead of the otherwise usual three-digit area code. This leads to an error, as a parenthesis cannot be converted to a whole number (Figure 7-6).

If you click the cell containing *Error*, you will see the error message at the bottom of the screen, like in Figure 7-6. If you directly click on the text *Error*, the query is filtered down to the error message alone.

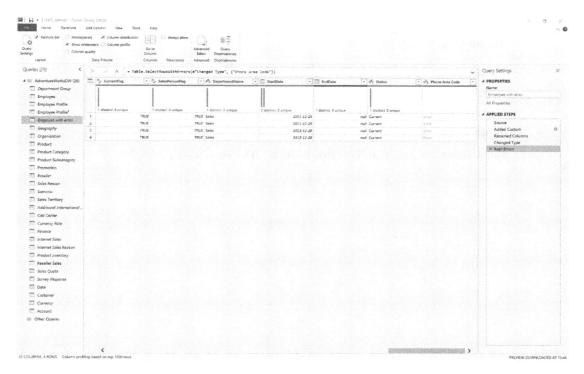

Figure 7-6. *Keep Errors in action*

- *Remove Duplicates*: Removes all rows where a certain value is contained in more than one row. Be careful, as you will lose details in other columns of the query (which share the same *BirthDate*). *BirthDate* will show 271 rows.

- *Remove Empty*: Filters out all rows where this column is empty. You use this filter on columns that necessarily contain a value. Be careful, as you will lose details in other columns of the query. Nothing happens for *BirthDate*, as there are no empty rows.

- *Remove Errors*: Removes all rows from the query where the current row has an error. You use this filter on columns that necessarily contain a valid value. Be careful, as you will lose details in other columns of the query. Doesn't change anything for the *BirthDate* column.

- *Replace Errors*: Replace all errors with a default value. Be careful, as the report user doesn't see any hint that there was an error, but sees the new (default) value only.

- *Show the column quality peek* (Figure 7-7): Shows a tooltip with absolute numbers and percentages (relative to the total amount of rows) of how many rows are *valid*, have an *error*, or are *empty* (for explanation, see section "Column Quality"). You can copy the exact same information into the clipboard with *Copy Quality Metrics*.

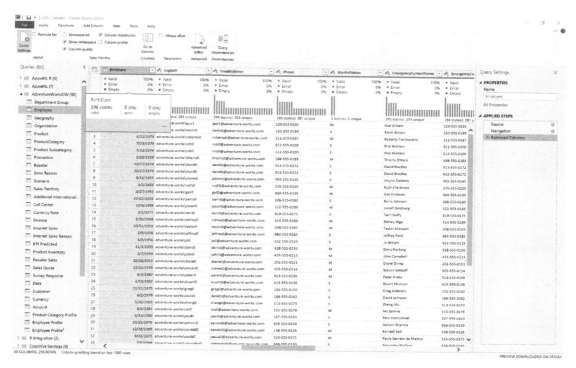

Figure 7-7. *The column quality peek for BirthDate*

- *Show the column distribution peek*: Shows a tooltip with absolute numbers and percentages (relative to the total number of rows) of how many rows are *distinct* and *unique* (see explanation in section "Column Distribution"). The absolute numbers are the same as those displayed underneath the column profile column chart. You can copy the exact same information into the clipboard with *Copy Value Distribution*. You can quickly select to remove duplicates (Figure 7-8).

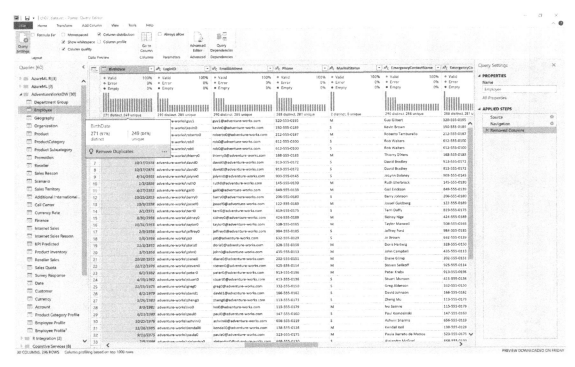

Figure 7-8. *The column distribution peek for BirthDate*

Copy Quality Metrics fills the clipboard with absolute numbers and percentages (relative to the total number of rows) and a categorization in *Valid, Error,* and *Empty.* For *BirthDate* I get the following result:

Valid	**296**	**100%**
Error	0	0%
Empty	0	0%

Copy Value Distribution fills the clipboard with a count of the number of rows per value in the column (for the biggest fifty values). Column BirthDate shows the following distribution:

1975-08-06	**3**
1988-02-04	3
1977-09-10	3

(*continued*)

1975-08-06	**3**
1980-06-13	2
1986-09-10	2
1986-05-01	2
1962-04-13	2
1974-10-17	2
1977-08-29	2
1989-12-15	2
1975-09-11	2
1990-06-01	2
1974-10-15	2
1986-04-19	2
1970-07-03	2
1975-11-15	2
1981-06-06	2
1986-04-10	2
1977-08-17	2
1984-07-29	2
1975-10-05	2
1974-07-23	2
1956-02-09	1
1985-12-28	1
1976-10-25	1
1955-10-31	1
1955-08-16	1
1988-07-05	1
1981-08-03	1

(continued)

1975-08-06	**3**
1956-03-30	1
1983-05-26	1
1975-08-18	1
1979-06-02	1
1971-03-01	1
1974-06-12	1
1978-02-15	1
1989-01-23	1
1958-10-09	1
1986-12-19	1
1968-04-17	1
1979-02-03	1
1972-06-25	1
1982-06-03	1
1982-10-31	1
1987-03-27	1
1970-10-16	1
1975-02-06	1
1974-08-11	1
1977-08-05	1
1975-04-30	1

The list is similar to the first two columns in the report I created (Figure 7-4).

Column Profile

Independent from *Column quality* or *Column distribution* we can enable *Column profile* (in *View*, section *Data Preview* in Power Query). If the column distribution does not appear on the bottom, please click on a column name on top (Figure 7-9).

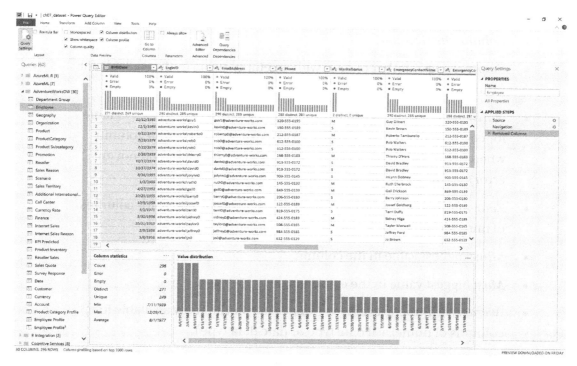

Figure 7-9. *The column profile for BirthDate*

This shows even more detailed profile information about the column than we got from either *Column quality* or *Column distribution*.

The information is split up into two sections:

- *Column statistics*

- *Value distribution*

The section *Column statistics* shows the following (depending on the data type of the column, not all statistics may show up):

- *Count*: total number of rows (as displayed at the bottom left)

- *Error*: number of rows with an error

- *Empty*: number of rows that are empty; those are the rows that show *null*

- *Distinct*: number of unique values (same number as in *Column distribution*)

171

- *Unique*: number of values that only appear once (same number as in *Column distribution*)

- *Empty string*: number of rows with an empty string (only available for columns of data type *Text*)

- *NaN*: number of rows where value is "Not a Number" (only available for columns of numeric data type). For example, a custom column where you divide 0 by 0 would show *NaN* as the result (and not an error).

- *Zero*: number of rows where value is zero (only available for columns of numeric data type)

- *Min*: smallest value in the column

- *Max*: biggest value in the column

- *Average*: arithmetic average of values in column (only available for columns of numerical data type or date data type)

- *Standard Deviation*: standard deviation of values in column (only available for columns of numerical data type)

- *Odd*: number of rows with an odd value (only available for columns with data type *Whole Number*)

- *Even*: number of rows with an even value (only available for columns with data type *Whole Number*):

- *True*: count of rows with value *True* (only available for columns of data type *True/False*)

- *False*: count of rows with value *False* (only available for columns of data type *True/False*)

The ellipses ("...") allows you to copy all the values into the clipboard.

Right from the *Column statistics* we get a column or bar chart (depending on the data type of the column) showing a histogram of the values. The ellipses ("...") allow you to either copy the values into the clipboard or to group the chart as follows:

- *Value*: shows the first fifty most common values; similar to the chart we get from *Column distribution*; is the default

- *Sign*: one column representing the number of positive values, one column representing the number of negative values (only available for columns of numeric data type)

- *Parity*: one column representing the number of even values, one column representing the number of odd values (only available for columns of numeric data type)

- *Text length*: one column per length of a text in that column (only available for columns of data type *Text*)

- *Year, Month, Day, Week of Year, Day of Week* (only available for columns of *Date* data type)

From the tooltip of each bar or column, we can directly start actions. Filter the rows in the current query on all columns either equal or not equal to the value in the chart. From the ellipses ("…") we have access to all filters. This filter applies to the whole query and will limit the rows loaded into Power BI's data model. If you do not want to apply the filter to the query, you can delete the action from the *Applied steps* in the *Query Settings* on the right to undo the action.

We can get the following interesting insights from the *Employee* query:

- The company seriously expanded its head count in year 2008, which we can see from a clear spike in *HireDate*, if grouped by year.

- *HireDate* has many values in months July, August, and September, some in October and December. The rest of the months have only one hire or none.

- *HireDate* is almost evenly distributed over the days of the week, with Saturday as the third most common day of the week. In some countries and/or industries it would be rather weird to hire someone on this day, as it is part of the weekend.

- With a single click on the histogram for *BirthDate* on value 1988-02-04 we learn who the three persons are who have their birthdays on the same day, month, and year.

- *Phone* numbers all have a length of twelve characters, except for four rows with a length of nineteen characters, which are prefixed with the international area code and a blank ("1 (11) "), as we learned in the section "Quality and Distribution Peek."

- *MaritalStatus* is evenly distributed (= 50:50) between married (*M*) and single (*S*). This is something you would probably not expect in real-world scenarios.

- Less than 20 percent of the employees have set *SalariedFlag* to *true*, meaning over 80 percent are paid by the hour.

- *Gender* distribution shows a binary gender with 70 percent males and 30 percent females.

- *DepartmentName* shows an overwhelming majority of employees in the production department on the one hand and only three employees in the executive department on the other.

We could gain a lot of insight even before loading the data into Power BI's data model and building a single visual.

Table Profile

What if you want to build Power BI reports on the numbers we just saw in *Column distribution*? Power Query's function `Table.Profile` is your friend. To apply this function on a query, we have to write a new query from scratch and make use of the Power Query Mashup language, or, for short, M.

In the ribbon in *Home* select *New Source* and choose *Blank query*. A new query with name *Query1* is created, and you have to write the first (and in the case of this example the only) line of code (don't forget the equal sign at the beginning of the line):

```
= Table.Profile(Employee)
```

After you press Enter (or select the checkmark to the left of the input field), the center of the screen fills with plenty of information. Please give the query a useful name (under which it is added to the data model in Power BI) by editing the *Name* field in *Query Settings* on the right of the screen. I chose `Employee Profile`.

Don't get confused: This newly created query contains *one row per column* of query *Employee*. We get so many rows as there are columns in query *Employee*. We always get the same number of columns, and these are the columns we get:

- *Column* (yes, there's indeed a column with name *Column*): name of the column in the query we used as the only parameter for function `Table.Profile`.

- *Min*: smallest value

- *Max*: biggest value

- *Average* (only available for columns of numerical data type or date)

- *StandardDeviation* (only available for columns of numerical data type or date)

- *Count*: number of rows in the table; we get the same value for all rows

- *NullCount*: number of rows either empty or null

- *DistinctCount*: number of distinct rows in that column

The information is a subset of what we saw in the section "Column Distribution."

Every time data in query *Employee* is changed during a data refresh, query *Employee Profile* will automatically be updated for you. It will always represent the current content of query *Employee*.

As soon as you choose either *Close & Apply* or *Apply* in the menu under *File*, the results of all changed queries are loaded into Power BI's data model and we can build a report on the (meta) data of table *Employee*. In Figure 7-10, I created a table visual for all the columns available in table *Employee Profile*.

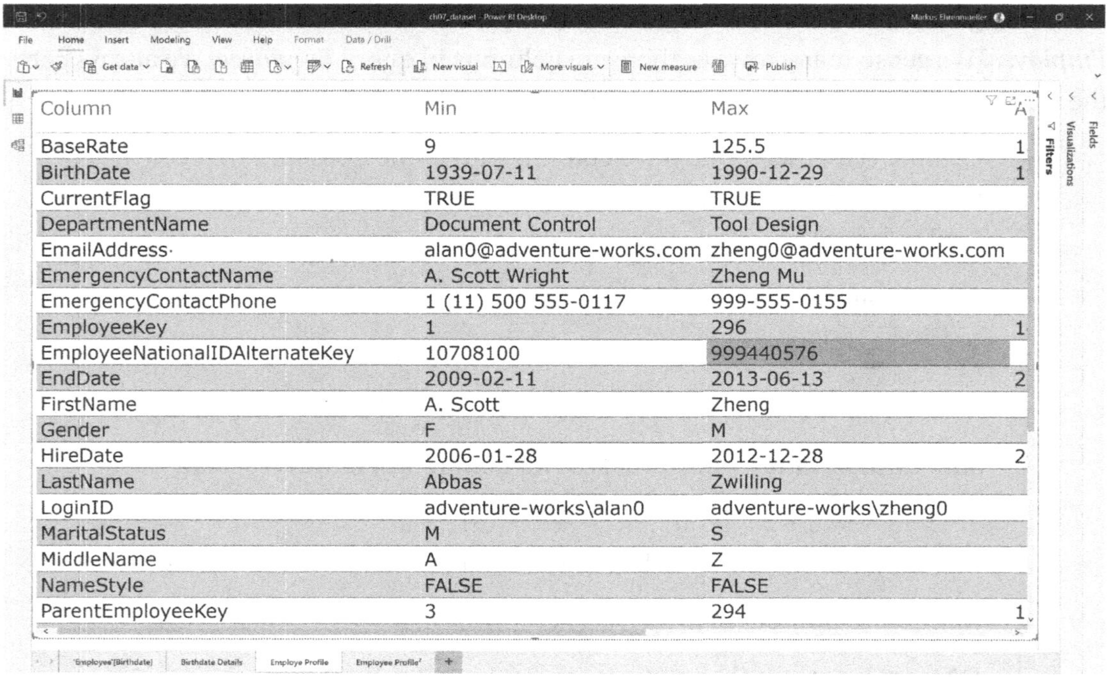

Column	Min	Max	
BaseRate	9	125.5	1
BirthDate	1939-07-11	1990-12-29	1
CurrentFlag	TRUE	TRUE	
DepartmentName	Document Control	Tool Design	
EmailAddress	alan0@adventure-works.com	zheng0@adventure-works.com	
EmergencyContactName	A. Scott Wright	Zheng Mu	
EmergencyContactPhone	1 (11) 500 555-0117	999-555-0155	
EmployeeKey	1	296	1
EmployeeNationalIDAlternateKey	10708100	999440576	
EndDate	2009-02-11	2013-06-13	2
FirstName	A. Scott	Zheng	
Gender	F	M	
HireDate	2006-01-28	2012-12-28	2
LastName	Abbas	Zwilling	
LoginID	adventure-works\alan0	adventure-works\zheng0	
MaritalStatus	M	S	
MiddleName	A	Z	
NameStyle	FALSE	FALSE	
ParentEmployeeKey	3	294	1

Figure 7-10. *Report on Employee Profile*

Such a report can be part of a data dictionary—no need for the report users to open Power Query first. In the Power BI data model, such a table will be unconnected; it does not make sense to have filters going from data (of, for example, table *Employee*) to metadata (e.g., *Employee Profile*) or the other way around.

Note Currently, there is no way to have profile data automatically generated for all queries. You have to create a query with `Table.Profile` explicitly for every single query, but you can append those queries to each other to get a single query / table containing the gathered profile metadata in the ribbon via *Home* ➤ *Append Queries* (in the *Combine* section).

If you have left *Column quality* and *Column distribution* turned on, we can analyze the content of *Employee Profile* (a.k.a. *Employee*'s metadata) in Power Query (Figure 7-11).

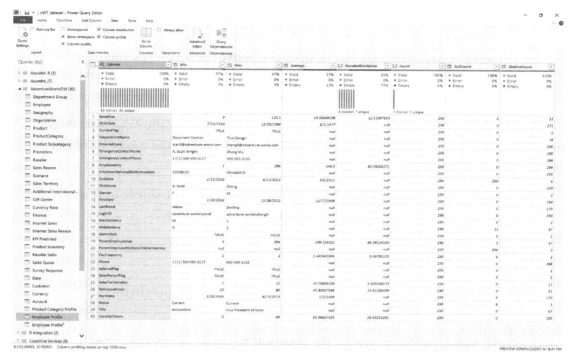

Figure 7-11. *Column quality and Column distribution for query Employee Profile*

- *Min* and *Max* are empty in 3 percent of the rows (a.k.a. columns of query *Employee*). That both are empty only happens if a column contains empty or null rows only, which is the case for exactly one column in query *Employee*: *ParentEmployeeNationalIDAlternateKey*. One out of thirty columns is 3 percent.

- *Average* contains valid rows (a.k.a. columns of query *Employee*) in 37 percent of the cases. We can conclude that slightly more than a third of the columns in query *Employee* are of type numeric or data (for which an average can be computed). I counted them for you: Seven columns are of data type *Whole Number* or *Fixed decimal number* and four are of data type *True/False*. Nine columns out of thirty is 37 percent.

- *StandardDeviation* contains valid rows (a.k.a. columns of query *Employee*) in 24 percent of the cases. We can conclude that slightly less than a quarter of columns in query *Employee* are of type numeric (for which a standard deviation can be computed). That's the seven columns with data type *Whole Number* or *Fixed decimal number*.

Do you want to drive this further? Calling `Table.Profile` for the *Employee Profile* query gives you metadata about metadata—that's why I named the query *Employee Profile*[2] (squared). There are only rare insights:

- Column *Count*, as it is always the same for all the rows, shows us that we have thirty columns in the original query (*Employee*).

- The names of the columns are alphabetically located between *BaseRate* and *VacationHours*.

- *Min* and *Max* for row *Count* shows us the number of rows in the original query (*Employee*).

- *StandardDeviation* was null in twenty-three cases (a.k.a. columns) and contains eight distinct values (with null as one of the distinct values). We knew this already: *Employee* contains seven columns with numeric data type, leaving twenty-three (thirty minus seven) with a non-numeric data type.

Key Takeaways

Power Query helps you to stay informed about the quality of your data as follows:

- Data quality is crucial for report quality. Features in Power Query help you to gain insights about your data and clean it up before you load it into Power BI's data model.

- *Column quality*: We get numbers of how many rows contain a *Valid* value, have an *Error*, or are *Empty*.

- *Column distribution*: We get the number of distinct values and the number of unique values in a column as well as a column chart showing the histogram.

The number of distinct values is the number of different values in the column. The number of unique values is the number of values that appear in only one row.

The difference between the row count and the number of distinct values gives us the number of rows that are not distinctly identified by this column value. This number gives us insight into the (values of the) rows of this column.

The difference between the number of distinct values and the number of unique values gives us the number of values that are used in more than one row. This gives us insight into the available content of this column.

- Column profile: Shows more statistical data (*Column Statistics*) and a histogram (*Value distribution*) we can group on different attributes.

- Function `Table.Profile` in Power Query's M language returns some of the values we get from *Column Statistics* as a query. We can load the content of the query into Power BI's data model and build reports on it.

## CHAPTER 8

Creating Columns from Examples

Power BI comes with a powerful tool (called Power Query) with which to import and clean data from different sources. The power of this tool derives from a powerful language: Power Query Mashup Language, or M for short. Instead of digging into the manual to learn the syntax of this language and finding the right functions to achieve your goal, you can transform data by just giving examples of what the new column should look like. A similar user experience is available for web scraping, if you want to load data from a web page. Everybody who likes Q&A to create new visuals (see first chapter of this book) will love the "Column from Examples" and "Web by Example" features. Before we use the smart way, let's take a look at the traditional way.

Power Query Mashup Language

In the ribbon in Power BI Desktop, select *Home* and *Transform data* (in the *Queries* section) to start the Power Query window. In Figure 8-1 I have selected query *Employee* on the left.

© Markus Ehrenmueller-Jensen 2020
M. Ehrenmueller-Jensen, *Self-Service AI with Power BI Desktop*, https://doi.org/10.1007/978-1-4842-6231-3_8

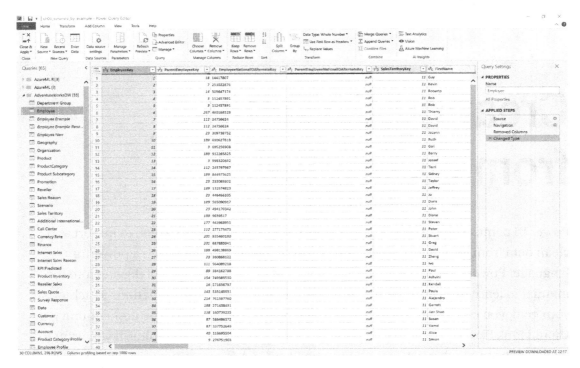

Figure 8-1. *Power Query window showing query Employee*

Every single transformation you make in Power Query is persisted as an *Applied Step* to the current query. All the steps are listed on the right side of the screen. If you click on one of the steps, you see the results of the query with all steps applied up to the selected one (but without the rest of the steps applied). This helps in debugging a query (e.g., when it does not deliver the expected result). In some cases, you can click on the gear symbol to reopen the wizard (or a similar window) with which you created the step originally. You can easily get rid of a step you don't need anymore by clicking the x icon in front of the step's name—but be careful, as this can't be undone (unless you close Power Query and discard any changes).

The names of the steps are automatically chosen (and numerated in the case of multiple steps with the same type of transformation) and give a hint of what kind of transformation you applied. Here is the list from Figure 8-2:

- *Source*
- *Navigation*
- *Removed Columns*

- *Changed Type*

- *Inserted Text Before Delimiter*

- *Inserted Day Name*

- *Added Custom Column*

- *Inserted Merged Column*

- *Added Conditional Column*

- *Inserted Merged Column1*

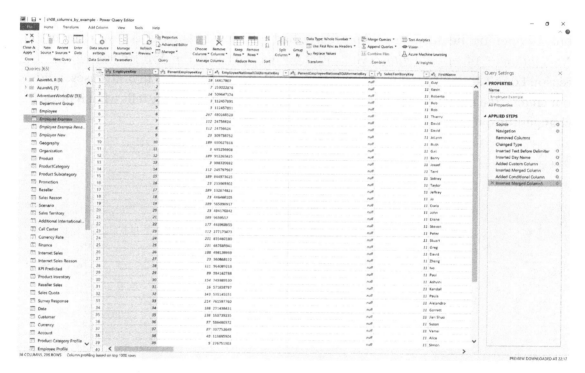

Figure 8-2. *Power Query Employee Example showing applied steps with generic names*

It's always a good idea to change the generic name of a step to describe in more detail what the step does (e.g., the name of the column created by that step). You can rename the step by either selecting it and pressing *F2* on the keyboard or right-clicking it and selecting *Rename*. Figure 8-3 shows the renamed steps from Figure 8-2, which are as follows:

- *Removed Columns ➤ Removed Photo and Linked Tables*

- *Inserted Text Before Delimiter ➤ Title without prefix*

- *Inserted Day Name ➤ Hire Day of Week Name*

- *Added Custom Column ➤ MiddleName with dot*

- *Inserted Merged Column ➤ FullName*

- *Added Conditional Column ➤ Salutation*

- *Inserted Merged Column1 ➤ Merged*

The first two steps (*Source* and *Navigation*) are special. You can rename *Source* only by opening the *Advanced Editor* (described in the next paragraph). If you rename *Navigation* in *Advanced Editor* it will still show *Navigation* in the *Applied Steps* for a reason I don't know. Therefore, I usually stick with the given names in both cases. I did not change the fourth step (*Changed Type*) either, because it already tells what it does: it changes the data types of the columns to the appropriate ones.

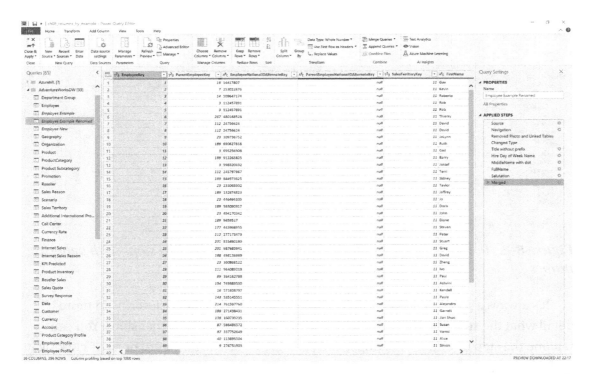

Figure 8-3. *Power Query window showing renamed Applied Steps in (duplicated) query Employee Example Renamed*

To see the M code behind a step, you've got two possibilities, as follows:

- Enable the *Formula Bar* (select *View* in the ribbon and then make sure that the checkmark at *Formula Bar* is ticked, as in Figure 8-4). This bar looks similar to the DAX formula bar in Power BI or the formula bar in Excel (but please don't confuse the syntax of Power Query M Language with DAX or Excel formulas). You can edit the selected step of the query directly in the bar. Click *x* (in front of the formula bar) to discard your changes, the checkmark to confirm your changes, or *fx* to insert a new step.

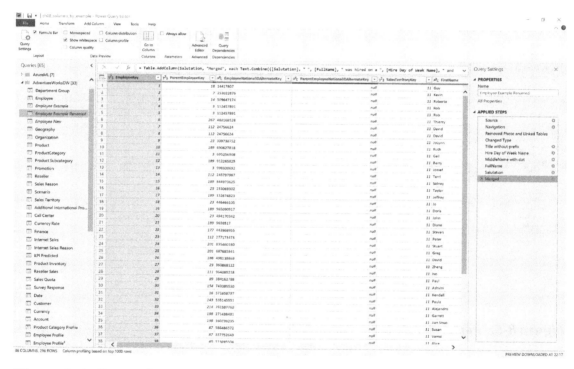

***Figure 8-4.** Power Query window with enabled Formula Bar*

- Open *Advanced Editor* (*View* ➤ *Advanced Editor* in the ribbon). The joke in the beginning days of Power Query was that it's called *Advanced Editor* because the user must have an advanced skill level to use it. This has changed for the better, but the editor is still not a full IDE (integrated development environment) similar to those you might recognize from other languages.

Every step listed in *Applied Steps* is one line of code in M. If the name of the step contains blanks or special characters, then the name of the step is masked with #" in the beginning and " at the end—making it unusual to read for beginners.

You can edit the whole query (all the steps) at once, which is useful for bulk edits. Sometimes I copy the whole query to a text editor and search and replace names of columns, which can be faster than editing single lines of code inside of Power Query (Figure 8-5).

Figure 8-5. *Power Query with Advanced Editor*

Add a Custom Column

The cases when you will actively write or change code in Power Query's Mashup Language will probably be rare, as the user interface (ribbon/menu bar) offers all available functionalities just a mouse click away. The most common situation where people use M is when they add a new column according to specific needs that are not offered (or immediately found) in the menu. Here follows the M code for the columns created in the *Applied Steps* discussed in query *Employee Example Renamed*:

- *Title without prefix*: Text.BeforeDelimiter([Title], " -") to get rid of the optional postfixes in column *Title* (e.g., "Production Supervisor" or "Production Technician"—without text "WC60" or "WC10" at the end)

- *Hire Day of Week Name*: Date.DayOfWeekName([HireDate]) to show the name of the weekday when the employee was hired (e.g., "Saturday")

- *MiddleName with dot*: Text.PadEnd([MiddleName], 2, ".") to add a dot after the middle name, if the middle name is available (e.g., "R.")

- *FullName*: Text.Combine({[FirstName], " ", [MiddleName with dot], " ", [LastName]}) to combine *FirstName, MiddleName with dot* and *LastName* (e.g., "Guy R. Gilbert" or "Roberto Tamburello")

- *Salutation*: if [Gender] = "M" then "Mr." else if [Gender] = "F" then "Mrs." else null to transform the gender into a salutation (e.g., "Mr." or "Mrs.")

- *Merged*: Text.Combine({[Salutation], " ", [FullName], " was hired on a ", [Hire Day of Week Name], " and had ", Text.From([VacationHours], "en-US"), " vacation hours so far."}) to combine some of the new columns into a full English sentence (e.g., "Mr. Guy R. Gilbert was hired on a Saturday and had 21 vacation hours so far.")

Column from Examples

Learning a new language is not easy, and Power Query Mashup Language is no exception, as you might have found out in the previous section. Fortunately, there is a way to create a new column without typing in M, by just telling Power Query how the result should look. In surprisingly many cases, Power Query will pick up what you are looking for and create the right formula in M for you.

To help Power Query detect the logic, I would recommend selecting the existing columns the new column should be based on (by clicking on the column headers while holding the Ctrl key on the keyboard, you can select multiple columns) and chose *Add Column* ➤ *Column from Examples* and *From Selection* (instead of *From All Columns*). In fact, all the examples from the previous section "Add a Custom Column" can be created as *Column from Examples*, as described in the following:

- In an example based on column *Title*, I entered `Production Technician` in the first line. All titles were then trimmed after the first two words. I double-clicked on `Senior Tool` in the fourth line and corrected it to `Senior Tool Designer`. It then picked up the correct logic. I named the column *Title without postfix* (after I double-clicked the column name) and pressed OK (Figure 8-6).

Figure 8-6. *Title without postfix*

- Getting the weekday for a given date (in my example: *HireDate*) is more simple than you might expect. If you double-click the first example, a whole list of different formats of a date is suggested: *Age, Day, Day of Week, Day of Week Name, Day of Year, Day in Month, Month, Month Name, Quarter of Year, Week of Month*, and several

start and end dates (of *Month, Quarter, Week,* or *Year*) are offered.
I just picked *Saturday (Day of Week Name from HireDate)* from the
list. You can see all available choices in Figure 8-7.

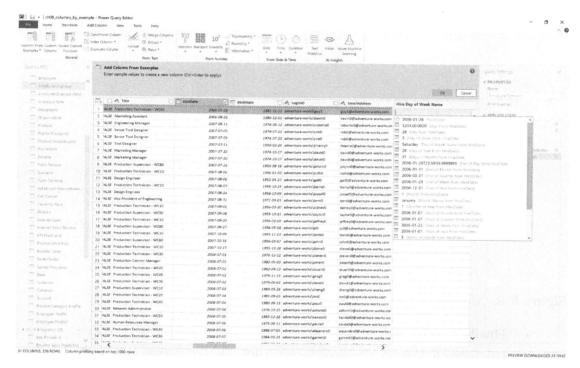

Figure 8-7. *Hire Day of Week Name*

- Selecting *FirstName, MiddleName,* and *LastName* and giving
  examples of a full name with the middle name abbreviated with a dot
  did not lead to the desired effect. Therefore, I split the task up into
  a) adding a dot to the middle name where appropriate (new column
  MiddleName with dot) and b) combined this column (instead of
  MiddleName) with *FirstName* and *LastName*.

For *MiddleName with dot* I gave only two examples: one for the first line (R.) and one
to correct Ann. (with a dot at the end) to Ann (without a dot at the end). You can see the
successful result in Figure 8-8.

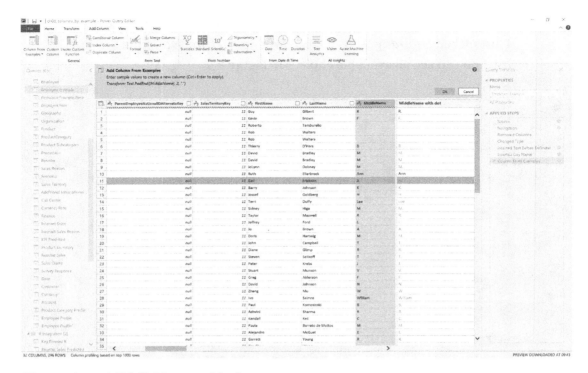

Figure 8-8. *MiddleName with dot*

- Finally, the *FullName* was simply based on the examples for the three columns *FirstName*, *MiddleName with dot*, and *LastName*, and I only had to give one single example until Power Query picked up what I was looking for (Figure 8-9).

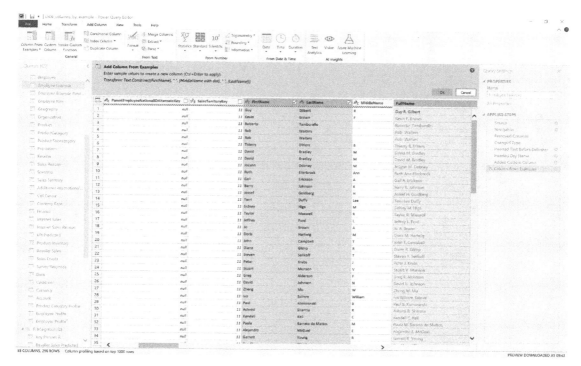

Figure 8-9. *FullName (based on FirstName, MiddleName with dot, and LastName)*

- To transform the one-character version of Gender ("F" or "M") to a salutation is simple: Enter Mrs. as an example for a row containing "F" and Mr. for a row containing "M" (Figure 8-10).

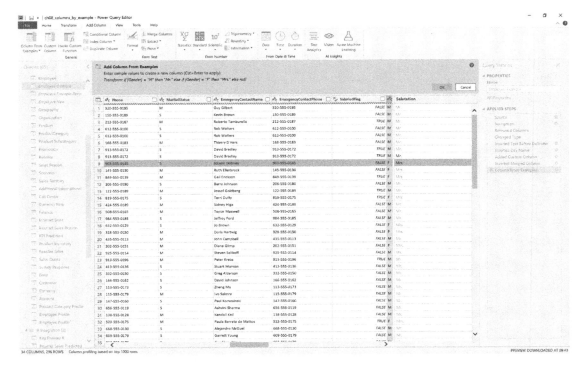

Figure 8-10. *Salutation*

- In the last example, I merged *Salutation, FullName, Hire Day of Week Name*, and *VacationHours* to a single English sentence. For the first row this was `Mr. Guy R. Gilbert was hired on a Saturday and had 21 vacation hours so far.` Lacking a good idea for a name of the column, I kept the generic name *Merged* (Figure 8-11).

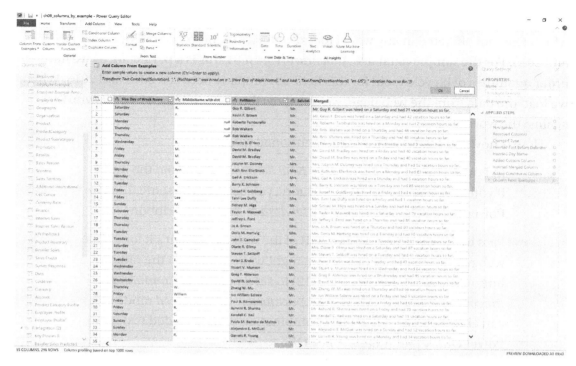

Figure 8-11. *Merged*

Attention   Please always double-check both the code and all the lines in the
query result to make sure that the formula will work in all cases, and not only
in the first few lines you can see. There is always a chance that Power Query
misunderstands the logic (e.g., instead of watching out for the first character after
a blank it will look out for the fifth character in the text because in all examples
provided the blank was in the fourth position).

Web Scraping

"Web scraping" is the term for using a web page as a data source. In Power Query, you
can *Get data* from *Web*. I will show you the classical approach first (where Power Query
analyzes the web page for you to discover data tables), and in the next section ("Web
by Example") I will show how you can provide examples of what you are interested in
(similar to the experience in the previous section about "Column from Examples").

At `https://www.nbda.com/articles/industry-overview-2015-pg34.htm` (with which I am not affiliated in any way) I found statistics on bicycles sold between 1981 and 2015. For the sample database *AdventureWorksDW* this is a perfect fit, as we have order quantities of bicycles sold for this fictious company between 2010 and 2013. When we load the data from the web page, we can show the performance of *AdventureWorks* against the world market.

Here is how you load data from the mentioned web page:

- In Power Query, select *Home* ➤ *New Source* ➤ *Web* from the menu. If you have not opened Power Query, you achieve the same thing in Power BI by selecting *Home* ➤ *Get data* ➤ *Web* in the ribbon.

- Pass in the URL (just mentioned) of the web page. That's it for public web pages. In special cases, you might want to click on the radio button *Advanced* (see Figure 8-12) to set an optional *Command timeout* or *HTTP request header parameter*. In *Advanced* mode you can also build the URL with a combination of hard-coded text and Power Query parameters. Click *OK*.

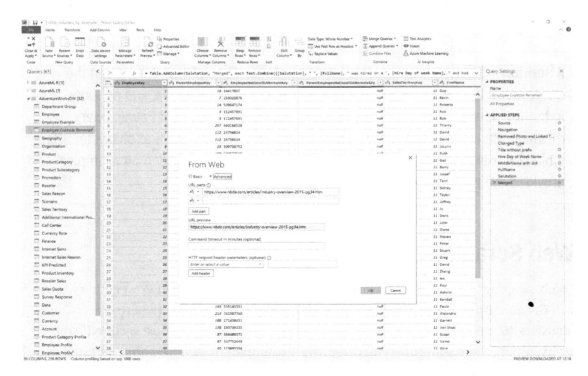

Figure 8-12. *Advanced option beside the basic URL*

194

- In the next step, you set the option for how you want to identify yourself against the web page. Public web pages can be accessed anonymously. In other cases, you have to ask the owner of the web page for the credentials and the necessary authentication mode. Click *Connect*.

  Your selection will be remembered by Power Query. If you need to change this choice later go to *File* ➤ *Options and settings* ➤ *Data source settings*, select the data source (i.e., the URL) and click *Edit Permissions* to directly change it. If you click *Clear Permission* instead, Power Query will forget the authentication setting. The next time you refresh from the data source or start a new query on this data source, the dialog from Figure 8-13 will appear again.

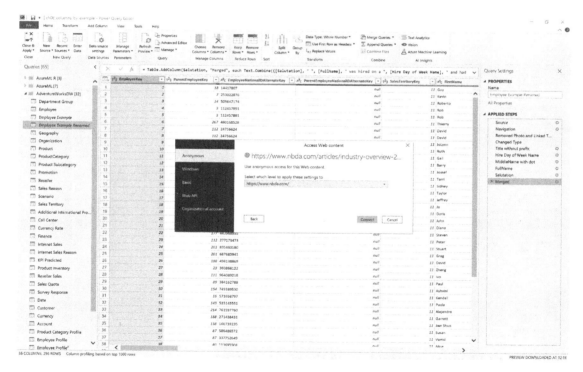

Figure 8-13. *Several options to identify yourself against the web content*

- Power Query is now connecting to the web page and looking for data in tabular form. In my case, it found three tables. It found two *HTML Tables* (generically named *Table 1* and *Table 2* by Power

Query given the lack of names in the HTML code) and one *Suggested Table* (named *Table 3*). If you click on the name of the table, you see a preview on the right of the window. The preview can be changed between *Table View* and *Web View*. *Table View* shows the data in a tabular form (which I prefer so as to find out if the table contains the data I was looking for and can be seen in Figure 8-14). *Web View* shows the content of the whole web page (which is helpful for orienting yourself about the overall content of the web page).

Table 1 contains statistics about worldwide bike sales I was looking for. Please tick the checkmark for *Table 1* and then click *OK*.

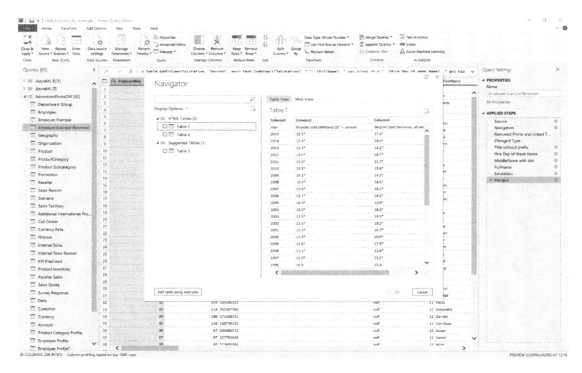

Figure 8-14. *Navigate to table with the right content*

- Back in the Power Query window, make sure to give the query an appropriate name (instead of the generic *Table 1*). I renamed it to *Bicycle Sales Worldwide* (Figure 8-15).

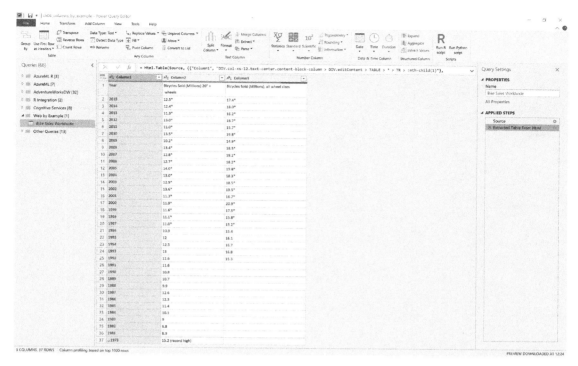

Figure 8-15. *Result for query Bicycle Sales Worldwide*

As soon as the data is loaded, there is no difference based on which data source supplied the data. All capabilities of Power Query (and Power BI) are available. To make the newly added *Bicycle Sales Worldwide* available in the data model, I would apply the following steps:

- *Transform* ➤ *Use First Row as Headers* as in Figure 8-16 (which adds step *Promoted Headers* into the list of *Applied Steps*)

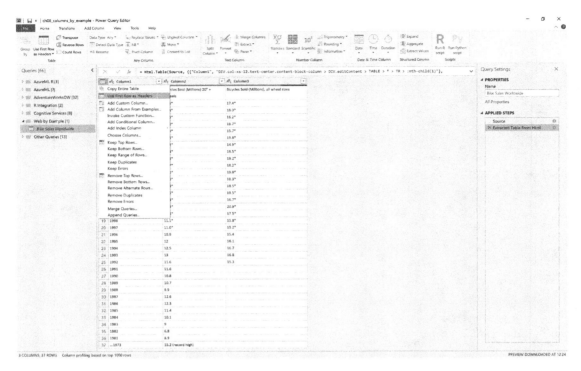

Figure 8-16. *Use First Row as Headers*

- Click the context menu (little triangle) of column *Year* and deselect
  ... 1973 (Figure 8-17).

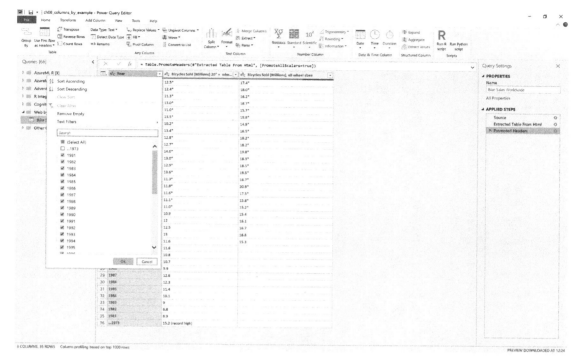

Figure 8-17. *Filter on column Year*

- Right-click on column *Bicycles Sold (Millions) 20"+ wheels* and select
  Remove (as we don't need these values), as shown in Figure 8-18.

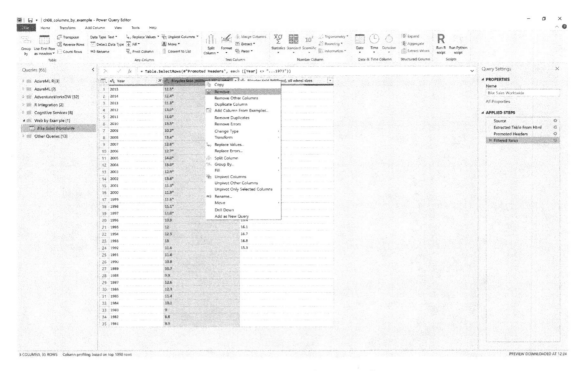

Figure 8-18. *Remove column Bicycles Sold (Millions) 20"+ wheels*

- Right-click on *Bicycles Sold (Millions), all wheel sizes* again and select *Replace Values*. Enter * in *Value To Find* and leave *Replace With* empty, as shown in Figure 8-19. Click *OK* to get rid of the asterisk.

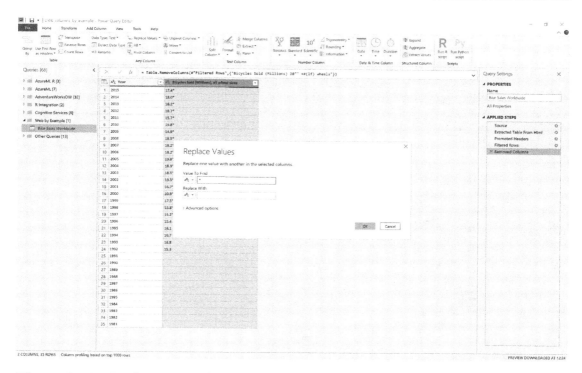

Figure 8-19. *Replace asterisk in column Bicycles Sold (Millions), all wheel sizes*

- Click on *ABC* to the left of the column name *Year* to change the data type to *Whole Number* (Figure 8-20). Do the same for the second column to change its data type to *Fixed decimal number*. In case you do not get *17.40* for *2015*, but *174*, make sure to either replace the decimal dot to a decimal comma in a step before or change *Local for import* in the *Regional Settings* for the *Current File* (under *File* ➤ *Options and Settings* ➤ *Options*) to one, where a dot is used as the decimal separator (e.g., *English (United States)*) and refresh your query to apply the new setting.

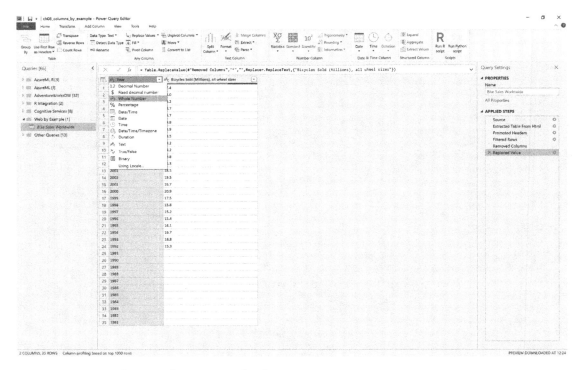

Figure 8-20. *Change data type of column Year*

- Select *Bicycles Sold (Millions), all wheel sizes* and choose
  Transform ➤ Standard (in the *Number Column* section) and
  Multiply (Figure 8-21). Enter 1000000 (one million) and click *OK*.

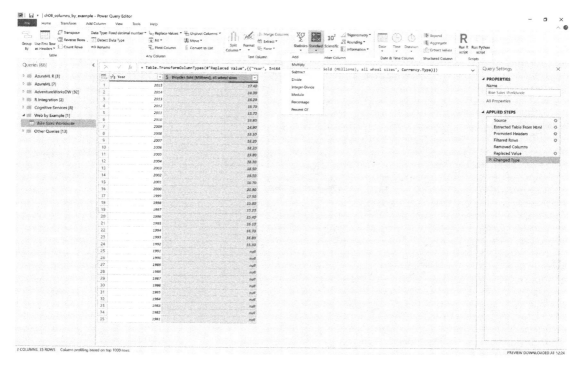

Figure 8-21. *Multiply the content of row Bicycles Sold (Millions), all wheel sizes*

- Rename column *Bicycles Sold (Millions), all wheel sizes* to
  Bicycles Sold Worldwide (by right-clicking the column name
  as in Figure 8-22).

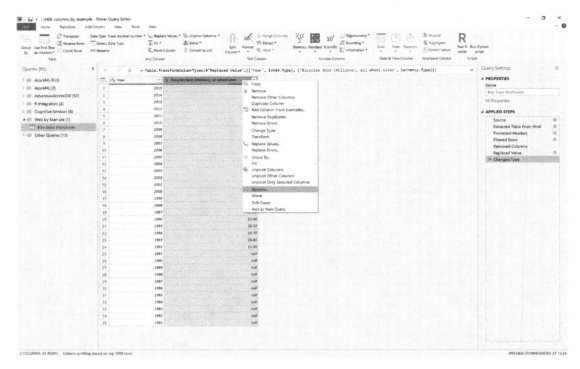

Figure 8-22. *Rename column to Bicycle Sales Worldwide*

- Optionally: Click on the triangle to the right of the column name for *Year* and apply the following *Number Filter*: *Greater than or equal to* year *2010*.

You see the final result of all the steps in Figure 8-23.

Figure 8-23. *Result for query Bicycle Sales Worldwide*

After you select *Home ➤ Close & Apply*, the new query will be added as a table to the Power BI data model. Don't forget to add a filter relationship between the *Year* column in table *Bicycle Sales Worldwide* and column *Calendar Year* in the *Date* table (in Power BI in the *Model* view; Figure 8-24).

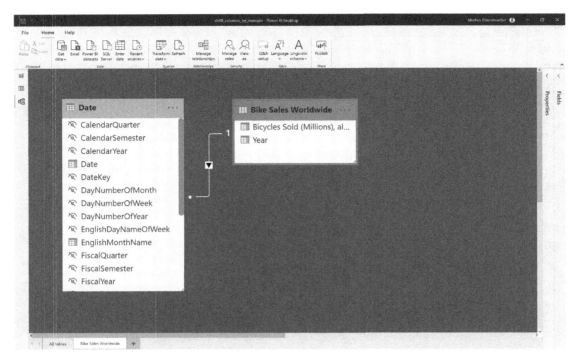

Figure 8-24. *Add a filter relationship between column CalendarYear in table Date and column Year in table Bicycle Sales Worldwide*

This is a special relationship. As we have several dates with the same *CalendarYear* in table *Date*, but only one row per *Year* in table *Bicycle Sales Worldwide*, the relationship is a many-to-one from *Date* to *Bicycle Sales Worldwide*. Therefore, the default filter direction will be that *Bicycle Sales Worldwide* will filter the *Date* table (as shown in Figure 8-24). This does not make any sense in our case, as we apply filters in the *Date* table, which will be forwarded to, for example, the table *Reseller Sales* and not the other way around. To make filters go from table *Date* to *Bicycle Sales Worldwide*, you must change the filter direction to *Both* in this case. Double-click the line representing the filter relationship and change the selection in *Cross-filter direction* from *Single* to *Both* (Figure 8-25). Please be careful with filter direction *Both* in other use cases, as you might end up with an ambiguous filter path in your model sooner or later.

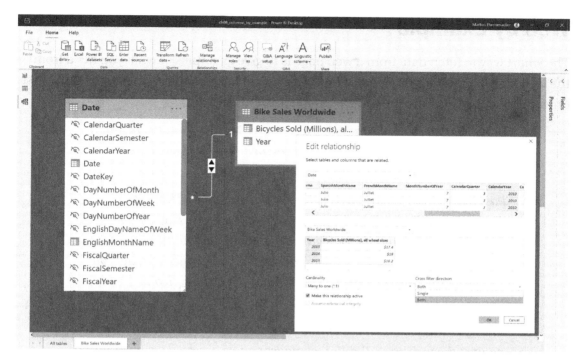

Figure 8-25. *Change Cross-filter direction to Both (only in this case)*

After that, you can start building reports showing both the bike sales of company *AdventureWorks* and the numbers from the world market.

Note   Merging data from different sources is one of the strengths of Power Query. But please be aware that external data sources, like a web page, might lack a service-level agreement. Nobody can or will guarantee you that the page will be available every time you want to refresh your Power BI data model. What is available publicly and for free now can change to a web page that asks you for a subscription key or be offline later. The structure of the web page can change anytime. Possibilities for changes are plentiful: name of URL or sub-pages, name of columns of HTML table, position and structure of HTML elements, etc. And any of those changes will probably break your Power Query (it will return an error instead of the desired query result).

This is true for our example as well. At time of writing, this example worked fine. But neither the author nor the publisher can guarantee that this example will still work at time of reading of this book.

Web by Example

The smart way to fetch data from a web page is to give examples of what you need from the web page (similar to the *Column from Examples* feature). Follow these steps (the first ones, printed in *italics*, are the same as in the previous section):

- *In Power Query select Home ➤ New Source ➤ Web from the menu. If you have not opened Power Query, you achieve the same in Power BI by selecting Home ➤ Get data ➤ Web in the ribbon.*

- *Pass in the URL (mentioned earlier) of the web page. That's it for public web pages. In special cases, you might want to click on the radio button Advanced (see Figure 8-12) to set an optional Command timeout or HTTP request header parameter. In Advanced mode, you can also build the URL with a combination of hard-coded text and Power Query parameters. Click OK.*

- *In the next step, you set the option for how you want to identify yourself against the web page. Public web pages can be accessed anonymously. In other cases, you have to ask the owner of the web page for the credentials and the necessary authentication mode. Click Connect.*

  Your selection will be remembered by Power Query. If you need to change this choice later go to File ➤ Options and settings ➤ Data source settings, select the data source (e.g., the URL) and click Edit Permissions to directly change it. If you click Clear Permission instead, Power Query will forget the authentication setting. The next time you refresh from the data source or start a new query on this data source, the dialog from Figure 8-13 will appear again.

- In the *Navigator* window (Figure 8-14) we ignore all options but click on button *Add table using examples* on the left bottom of the window.

- Scroll the preview of the web page down to the section you are interested in (the table containing column *Bicycles Sold (Millions), all wheel sizes*) to orient yourself. I then entered 2015 and 2014 in the empty rows underneath *Column1*. I then got a message that *No CSS selector was found for the sample values you provided in the following column(s): Column1. However, moving your sample values would allow us to find a match. Would you like us to move them for you?* (Figure 8-26).

This happens because the inputted values 2015 and 2014 appear on different sections of the web page, but Power Query is smart enough to match them to the rows in the table we want to load. Click on *Move* to see with your own eyes: *Column1* now got filled with a) the header (*Year*) and all the years (we make *Year* the column name in a later step).

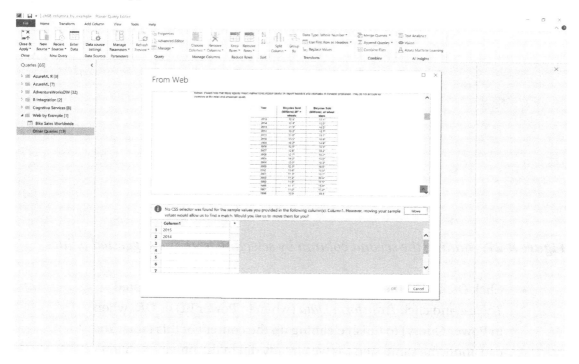

Figure 8-26. *No CSS selector was found, but the wizard is here to help you*

- In the right-hand column (headed with an asterisk) double-click and start writing `Bicycle` and then select `Bicycles Sold (Millions)`, `all wheel sizes` (Figure 8-27). As soon as you press Enter on the keyboard, you get the correct numbers brought into the table (with the header in the first line, which we solve in a later step). Please refrain from renaming the column headers in the first line or changing the format of the content (e.g., 17400000 instead of 17.4*); apply those transformations in later steps in Power Query. We do not want to overwhelm *Web by Example* with transformations. Use it to load the data as it is.

Figure 8-27. *Adding the second column by selecting from the suggested values*

- Click *OK* and in the *Navigator* window select the newly created
  Table 4 and click *Transform Data* (when in Power BI) or *OK* (when
  in Power Query) to finish cleaning up the content of this table. You
  can apply the same steps as we already did in the previous chapter
  (cleaning the columns, loading the query as table, and creating a
  filter relationship in the Power BI model, as described in the section
  "Web Scraping").

In the preceding example, we easily achieved the same result with a comparable
similar effort. Depending on the complexity of the web page's code, of course, your
mileage may vary. That means I had real-world cases where the simple web data did not
lead to any useful way of loading the content of the web page, because no (HTML) table
was found, but Web by Example helped me to connect the pieces "spread" out in the web
page's code.

Key Takeaways

Power Query is a powerful tool for the following reasons:

- Power Query Mashup Language, or M, is the language used to apply transformations in Power Query. You can write your own scripts in M step-by-step or in *Advanced Editor*. Or you can apply the steps via the ribbon (which will then generate the M code for you). All steps are stored with the query and will be reapplied every time you refresh the data.

- *Web scraping* is the term for loading data from a web page. In Power Query you use the data source named *Web* for that.

- *Column from Examples* allows you to write concrete examples of what you need, and Power Query will come up with the right M code to transform the other column(s) into that.

- *Web by Example* allows you to write concrete examples of what you need, and Power Query will come up with the right M code to transform the content of the web page into that.

## CHAPTER 9

Executing R and Python Visualizations

If Power BI's standard or custom visuals cannot satisfy your need to present the data in a certain way, or if you already have a data visualization script written in either R or Python at hand, then the R visualization and Python visualization are your friends. You need a local installation of R or Python, with all the packages you want to use, and you can then copy and paste your script into the Power BI visualization. And there we go with a new, powerful visualization in Power BI.

R and Python

Both languages are very popular among data scientists. It seems like you can't learn anything about machine learning without learning at least one of these two languages. Most people I have met so far do prefer one (highly) above the other, though (to articulate this fact as politely as possible). Both languages (and packages) are open source and free to use. You will find plenty of books about R and Python, from beginner to advanced levels.

While I assume that you, dear reader, already have some experience with R and/or Python, the examples are as simple as possible. I tried to come up with a working example without introducing too much code or too many parameters. Don't consider them as perfect code to run in production, but rather as a starting point to give you ideas about what you can do. The primary goal for this book is to show what you can achieve with Power BI Desktop, and not so much what you can achieve with R or Python.

© Markus Ehrenmueller-Jensen 2020
M. Ehrenmueller-Jensen, *Self-Service AI with Power BI Desktop*, https://doi.org/10.1007/978-1-4842-6231-3_9

Both languages have the following in common:

- Strong capabilities to work with data and to implement machine learning (ML) algorithms through use of easy-to-install add-ins (so-called packages or libraries)

- Open source, with an active community driving the language and the packages forward

- Free to use—the perfect fit with Power BI Desktop, which is free to use as well

- Interpreted languages (like the other languages of business intelligence: DAX, M, SQL, MDX)

- R and Python are case sensitive (opposed to DAX, SQL, and MDX). If you receive an error message, see if it is due to the fact that you misspelled the case (e.g., `as.Date()` vs. `as.date()` in R).

I picked some of the visualizations from Chapter 5 ("Adding Smart Visualizations") and rebuilt them in both R and Python. But before we can play around in Power BI, make sure that you have installed and configured R and Python, as described in the following section.

Getting Power BI Ready for R

Even though neither this chapter nor this book is an introduction to R, I will guide you through the steps to run your (maybe?) first R script in Power BI.

Here is a checklist of things you have to prepare before you can run R code in Power BI:

- Download the necessary file according to the description on either `https://mran.microsoft.com/open` (Microsoft's enhanced distribution of R) or `https://cran.r-project.org`. At time of writing this chapter, the most recent version of Microsoft R Open was 3.4.4 (with which I created all examples).

- Install an integrated development environment (IDE). This is not mandatory, but is highly recommended so as to make developing and testing your code much easier. You will find different IDEs for R. I use R Studio (https://rstudio.com).

- For the examples in this book, you will need to install additional packages. Please execute the following statements in R Studio or other IDE:

```
install.packages("gridExtra", dependencies=TRUE)
install.packages("ggplot2", dependencies=TRUE)
install.packages("scales", dependencies=TRUE)
install.packages("dplyr", dependencies=TRUE)
install.packages("forecast", dependencies=TRUE)
install.packages("ggfortify", dependencies=TRUE)
install.packages("zoo", dependencies=TRUE)
install.packages("corrplot", dependencies=TRUE)
install.packages("tm", dependencies=TRUE)
install.packages("wordcloud", dependencies=TRUE)
```

- In Power BI's ribbon select *File* ➤ *Options and settings* ➤ *Options* ➤ *R scripting* (in the *Global* section; see Figure 9-1) and make sure that the correct *R home directory* and *R IDE* are selected.

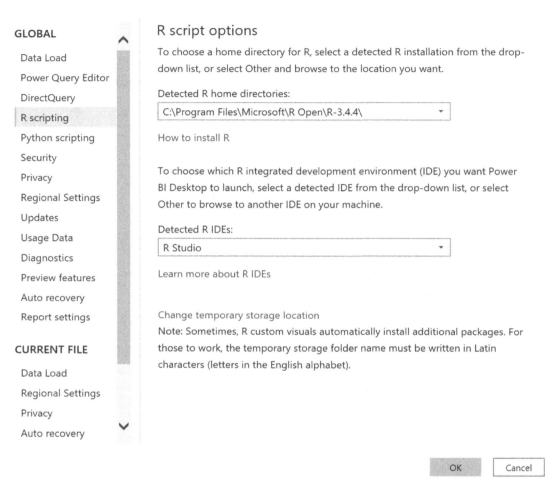

Figure 9-1. *Options for R scripting*

Getting Power BI Ready for Python

Even though neither this chapter nor this book is an introduction to Python, I will guide you through the steps to run your (maybe?) first Python script in Power BI.

Here is a checklist of things you have to prepare before you can run Python code in Power BI:

- Download the necessary file according to the description on https://www.python.org. At time of writing this chapter, the most recent version was 3.8.2 (with which I created all examples).

- Install an integrated development environment (IDE). This is not mandatory, but is highly recommended so as to make developing and testing your code much easier. You will find many different IDEs for Python. I use Visual Studio Code (https://code.visualstudio.com/) with the Python extension (https://marketplace.visualstudio.com/items?itemName=ms-python.python).

- Power BI needs two Python packages: *pandas* and *matplotlib*. Simply type pip install pandas and pip install matplotlib in the command shell of your computer to download and install those packages. Both packages are automatically loaded when the Python script visual is executed.

- For the examples in this book, you will need more packages. Also execute the following:

```
pip install numpy
pip install scikit-learn
pip install statsmodelspip install seaborn
pip install wordcloud
```

  Some of these packages might ask you to install other tools (e.g., Visual C++). Read the (error) messages carefully to find out what is missing.

- In Power BI's ribbon, select *File* ➤ *Options and settings* ➤ *Options* ➤ *Python scripting* (in the *Global* section; Figure 9-2) and make sure that the correct *Python home directory* and *Python IDE* are selected.

Options

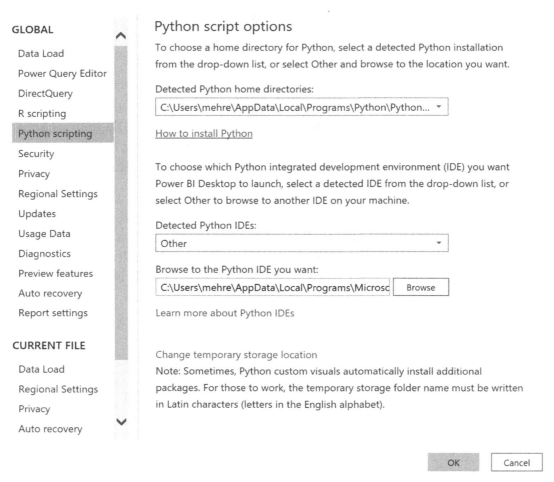

Figure 9-2. *Options for Python scripting*

Introduction to R and Python Visualizations

You can find the *R script visual* and the *Python script visual* as standard visualizations available in every Power BI Desktop file in the *Visualizations* section on the right.

If you start with a new R script visual and add fields into the *Values* section, the new script will be empty, except for the comment lines shown in Figure 9-3.

```
# The following code to create a dataframe and remove duplicated rows is
always executed and acts as a preamble for your script:

# dataset <- data.frame(Date, Sales Amount)
# dataset <- unique(dataset)

# Paste or type your script code here:
```

Figure 9-3. *A new R script visual*

If you start with a new Python script visual and add fields into the *Values* section, the new script will be empty, except for the comment lines shown in Figure 9-4.

```
# The following code to create a dataframe and remove duplicated rows is
always executed and acts as a preamble for your script:

# dataset = pandas.DataFrame(Date, Sales Amount)
# dataset = dataset.drop_duplicates()

# Paste or type your script code here:
```

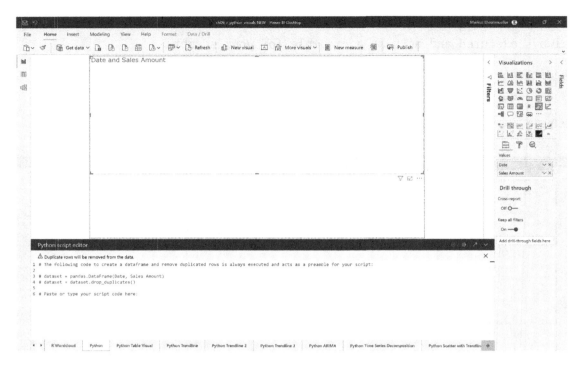

Figure 9-4. *A new Python script visual*

As those lines are comments only, you can remove them without any effect on the functionality of the script. Changing the comment (e.g., renaming data frame *dataset* to a different one or updating the list of columns) will have no effect either.

The automatic comments are just reminders that the columns and measures you have selected as *Values* for the visual are injected into the R/Python script in a data frame with the name *dataset* (in the example code I selected column *Date* and measure *Sales Amount*). Important: Power BI is only exposing a list of unique values.

Keep in mind that, additionally, Power BI will always automatically group and aggregate the values. For example, if you select measures only, data frame *dataset* will contain a single row with the aggregated values. Or if you select the *Year* column and measures, data frame *dataset* will only contain a single row per year with the aggregated values. You can avoid the automatic aggregations either by using calculated columns instead of measures and setting *Aggregation* to *Do not summarize* (to get single values instead of aggregates) or by adding a column to the *Values* that contains a distinct value on the needed granularity level (i.e., a unique identifier like a date, product name, or order number or one you have created in Power Query). Back in Chapter 3 ("Discovering

Key Influencers") we already discussed the importance of choosing the right granularity and how to create an index column if you don't have a row identifier in your table.

No matter if you use scripts in R or in Python, all the scripts more or less contain the following steps (single steps might be omitted):

- Load packages/libraries

- Prepare dataset

- Create model

- Create plot

The last step is mandatory. The script must create a plot, otherwise the visual cannot show anything (and will show an error message instead). In the case of Python, this is most likely due to a forgotten `plt.show()`.

Simple R Script Visual

The first R script visual I have chosen to show you plots measure *Sales Amount* over column *Date* as a scatter plot. It is a very simple script that contains two lines of code only (make sure to add both columns as fields in the *Values* section).

The first line converts the content of the *Date* column (from the format Power BI Desktop provides in the data frame of name *dataset*) into a format more usable with most packages I work with in R. *POSIXct* is the so-called UNIX date/time format. It represents a point in time as seconds since January 1, 1970. If you omit this step, the script will still work, but the x-axis will show the dates in a different format.

```
# prepare dataset
dataset$Date <- as.POSIXct(dataset$Date)
```

The second, and in this case final, line, passes the *dataset* to the `plot` function of R's base package. For getting a first impression of data, the `plot` function is sufficient. It shows a scatter plot by default. Add `type="l"` as a parameter to get a line chart, for example. In later sections, we will build more sophisticated charts with the help of the `ggplot2` package.

```
# create plot
plot(dataset)
```

In Figure 9-5, you can see the whole script and the result.

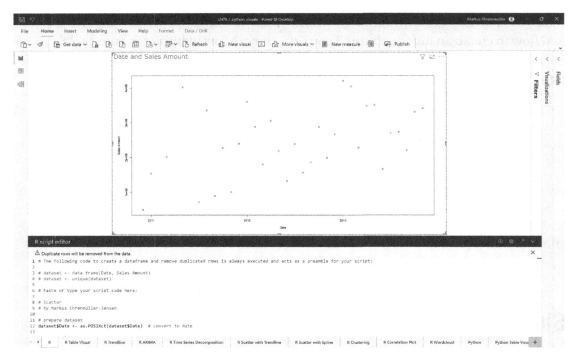

Figure 9-5. *Scatter plot in R*

Simple Python Script Visual

The first Python script visual I have chosen to show you is similar to the R visual in the previous section: plotting *Sales Amount* over *Date* as a line plot. It is a similarly simple script and contains three lines of code only.

The first line converts the content of the *Date* column (from the format Power BI Desktop provides in the data frame of name *dataset*) into a more usable format. If you omit this step, the script will still work, but the x-axis will show the dates in a different format (making it hard to read because of overlapping).

```
# prepare dataset
dataset.Date = pandas.to_datetime(dataset.Date)
```

The second line calls the `plot` method for the data frame *dataset*. For getting a first impression of data, the `plot` method is sufficient. It shows a line chart by default. Add `kind="scatter"` as a parameter to get a scatter plot, for example. In later sections, we will build more sophisticated charts.

```
# create plot
dataset.plot(x='Date', y='Sales Amount')
```

Creating the plot is not enough in Python (as opposed to R). You need to explicitly show the plot by calling the function show (which is part of the *matplotlib* package).

```
matplotlib.pyplot.show()
```

In Figure 9-6, you can see the whole script and the result.

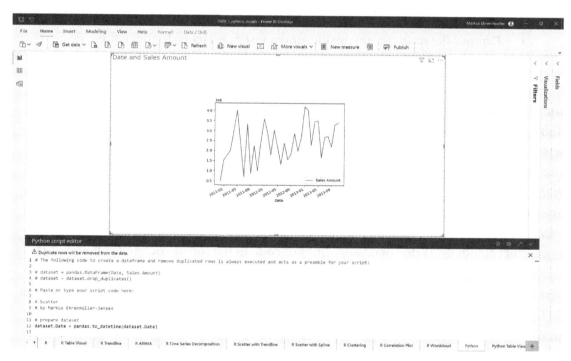

Figure 9-6. *Line plot in Python*

R Script Editor and Python Script Editor

Both the R script editor and the Python script editor offer the following functional buttons in the top right header bar (as seen in Figure 9-6):

- *Run script* (a triangular icon) to re-run the script. This is helpful after you change the script.

223

- *Script options* (gear icon) opens the same dialog as in *File* ➤ *Options and settings* ➤ *Options* ➤ *R scripting* or *Python scripting* respectively (see sections "Get Power BI Ready for R" and "Get Power BI Ready for Python").

- *Edit script in external IDE* (arrow pointing to left top) is really helpful (see following paragraphs to read why).

- *Minimize the script pane* (v-shaped icon) to see the whole report pane; is only available when the script pane is expanded.

- *Expand the script pane* (upside-down v-shaped icon) to make the script visible; is only available when the script pane is minimized.

Editing scripts in the editor inside of Power BI Desktop comes with the downside that you can only show data in the form of a single plot. The sections "R Script Visual Table" and "Python Script Visual Table" demonstrate how you can visualize the data as a table instead of a chart. When I want to debug my script, I prefer to print the values of data frames and other variables, though. The name of the feature *Edit script in external IDE* assumes that the script will be simply opened in an external editor (the one you specified in the R script options or Python script options). But it does a little more: Power BI automatically exports the data you selected for this visual as a CSV file into a temporary folder and adds necessary lines of code to the top of your script before it is opened in the external IDE. This way you can fully write, test, debug, and finish your script on the exact data available in Power BI before you copy and paste it back to Power BI Desktop. Don't forget to remove the automatically added lines, though. This feature makes the combination of Power BI Desktop and R/Python very handy. Extracting and transforming data inside of Power BI Desktop (via Power Query) can be more convenient than writing a script in R or Python doing the same job. *Edit script in external IDE* can therefore be an interesting feature, even if you are not planning to build a visual in Power BI, but continue to work in the IDE instead.

As you can see in Figure 9-7, in the case of an R script, *Edit script in external IDE* opens R Studio (as this is the editor I chose in the options). The first three lines are automatically added to the script (to load the data temporarily exported by Power BI Desktop into data frame *dataset*):

```
# Input load. Please do not change #
`dataset` = read.csv('C:/Users/mehre/REditorWrapper_c7301965-1d8d-4b2b-
aa28-ec4950ca6a83/input_df_6596a922-5c98-498e-9082-578236680a63.csv',
check.names = FALSE, encoding = "UTF-8", blank.lines.skip = FALSE);
# Original Script. Please update your script content here and once
completed copy below section back to the original editing window #
```

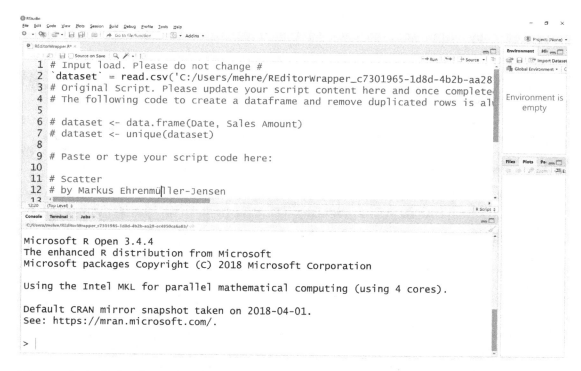

Figure 9-7. *R Studio with the script from the R script visual*

Edit script in external IDE opens Visual Studio Code in the case of a Python script, as you see in Figure 9-8 (as this is the editor I chose in the options). The first twelve (and the last two) lines are automatically added to the script (to load the data temporarily exported by Power BI Desktop into data frame *dataset*):

```
# Prolog - Auto Generated #
import os, uuid, matplotlib
matplotlib.use('Agg')
import matplotlib.pyplot
import pandas
```

```
os.chdir(u'C:/Users/mehre/PythonEditorWrapper_46cdc6f7-7417-44b6-9b8a-
9891680b48e9')
dataset = pandas.read_csv('input_df_e5c8eb59-7068-45c5-a824-885a5d842f76.csv')

matplotlib.pyplot.figure(figsize=(5.55555555555556,4.16666666666667), dpi=72)
matplotlib.pyplot.show = lambda args=None,kw=None: matplotlib.pyplot.
savefig(str(uuid.uuid1()))
# Original Script. Please update your script content here and once
completed copy below section back to the original editing window #
```

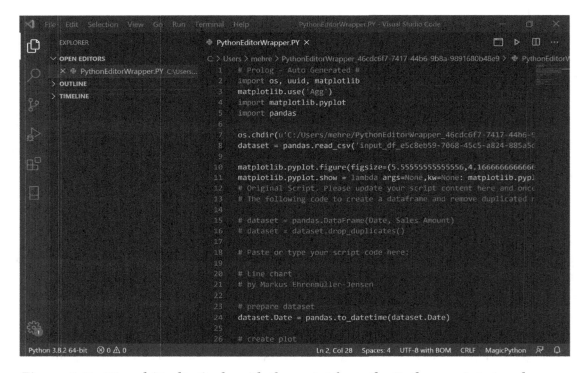

Figure 9-8. *Visual Studio Code with the script from the Python script visual*

R Script Visual: Table

This script is similar to and as simple as the one with the plot function from the base package. It loads library gridExtra and wraps the call to function tableGrob with grid. arrange. I use it sometimes as a poor man's debugger of the script running in Power BI.

```
# load package
library(gridExtra)
```

```
# create plot
grid.arrange(tableGrob(dataset))
```

Find the outcome in Figure 9-9.

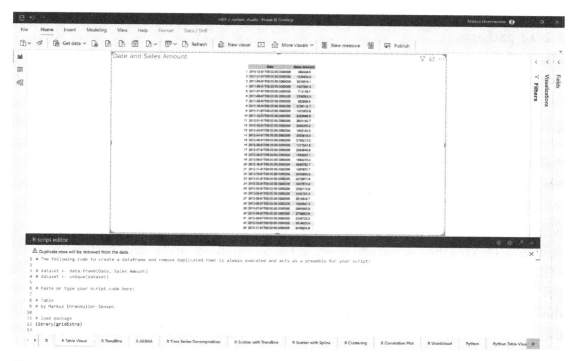

Figure 9-9. *R script visual showing a table*

I put all the other R scripts in the immediate next sections. Feel free to skip those sections in favor of concentrating your reading experience on the Python scripts.

R Script Visual: Trendline

Here comes the first R script visual that rebuilds one of the examples we saw in Chapter 5 ("Adding Smart Visualizations"): a trendline for the actual *Sales Amount* over *Date* (which I put into the *Values* section of the R script visual.)

First, I load the necessary libraries. *ggplot2* is a very common package for plotting charts. The first part of the package's name (*gg*) stands for grammar of graphics, which is a concept describing the chart in parts and in layers. The postfix (*2*) is indeed the hint that this package is the successor of package *ggplot*, which is not available anymore. Package *scales* allows you to print numbers in way business users will expect them (with thousands separator) instead of showing the number in scientific notation (e.g., 1E06 for one million).

```
# load packages
library(ggplot2)
library(scales)
```

Next, I convert the data type of column *Date* (as described in section "Simple R Script Visual"):

```
# prepare dataset
dataset$Date <- as.POSIXct(dataset$Date)
```

Finally, we create the plot with the `ggplot` function of *ggplot2*'s library. The first parameter is the *dataset* containing the data we want to plot. Then, we describe the aesthetics (`aes`) in the form of what we want to plot on the x- and y-axis. As the name of the measure *Sales Amount* has a blank in its name, I had to wrap it into function get. This was not necessary for column *Date*, which I used directly. geom_line plots the data in a line chart. `color=3` sets the color of the line to green. geom_point draws the data points (on top of the lines). Via `theme` I set the text size to 18 points, making it more convenient to read everything. `scale_y_continuous` limits the range of the y-axis (to show only values between 0 and 5m) and sets the number format to `comma` (a non-scientific format). In `theme` the `legend` gets disabled by setting its position to *none*. With `labs` I set the labels for both the x- and y-axis. That was all to create a line chart. The last line (`stat_smooth`) generates a linear model (a.k.a. linear regression) and adds it as an extra line. If you remove parameter `se` or set it to TRUE you will get a band showing the confidence interval.

```
# create plot
ggplot(dataset,
       aes(x=Date,
           y=get("Sales Amount"))) +
```

```
geom_line(color = 3) +
geom_point(color = 3) +
theme(text = element_text(size = 18)) +
scale_y_continuous(limits = c(0, 5000000), labels = comma) +
theme(legend.position = "none") +
labs(x = "Date",
     y = "Sales Amount") +
stat_smooth(method=lm, se=FALSE)
```

Parts of the code and the final visual are shown in Figure 9-10. Over the course of all three years we have a clear upward trend.

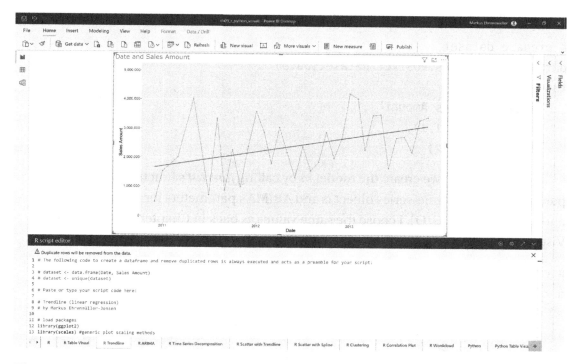

Figure 9-10. *R script visual showing a trendline*

R Script Visual: ARIMA

As described in Chapter 5 ("Adding Smart Visualizations"), ARIMA is the acronym for Auto-regressive Integrated Moving Average and is a machine learning model to generate time-series forecasts that can detect not only the trend but seasonality as well.

First, we load all necessary packages.

```
# load packages
library(zoo)
library(forecast)
library(dplyr)
library(ggplot2)
library(scales)
```

Column *Date* is converted into the POSIXct format and a time-series object with name *ts* is generated with the help of *zoo*'s ts function. The time-series' values are the *Sales Amount*, only data from January 2011 onward shall be contained, and the *dataset* contains monthly values (giving us the frequency of 12).

```
# prepare dataset
dataset$Date <- as.POSIXct(dataset$Date)
ts <- ts(
    dataset["Sales Amount"],
    start = c(2011,1),
    frequency = 12)
```

In the next step, we create the model *m* by calling *forecast*'s function Arima. The parameters are the time-series object *ts* and ARIMA's parameters for order (p, d, q) and seasonality (P, D, Q). I chose the same values as back in Chapter 5 ("Adding Smart Visualizations").

```
# create model
m <- Arima(
    ts,
    order = c(0, 1, 0),
    seasonal = c(0, 1, 0))
```

I pass in the model *m* and the number of months to *forecast*'s function predict and save the result in object *m.fit*.

```
m.fit <- predict(m, n.ahead = 6)
```

Then, I use a nested call to different functions to convert the content of *m.fit* into a data frame with name *m.fit.df* containing the dates and predicted values (*SalesAmountPred*) for the future and an empty (NA) column *SalesAmount*.

```
m.fit.df <- data.frame(
  Date=
    strptime(
      as.POSIXct(
        as.Date(
          as.yearmon(
            time(m.fit$pred)
          )
        )
      ),
      format = "%Y-%m-%d"
    ),
  SalesAmount=NA,
  SalesAmountPred=as.matrix(m.fit$pred)
)
```

Here, I merge (function from package *dplyr*) the content of *m.fit.df* (predicted future values) with the actual values (*dataset*). all = TRUE guarantees that I keep all rows from both data frames, even for *Date*'s available in only one of the data frames (people working with databases call this a full outer join):

```
dataset <- merge(
  x = dataset[,c("Date", "Sales Amount")],
  y = m.fit.df[,c("Date", "SalesAmountPred")],
  by = "Date",
  all = TRUE)
```

Finally, I plot *Date* and *Sales Amount* like in the script in the previous section, with the following differences: The limits of y-axis are extended to 6m (to fit the predicted values). And I add another geom_line and geom_point to show the predicted values (*dataset$SalesAmountPred*) in a different color.

```
# create plot
ggplot(dataset,
       aes(x=Date,
           y=get("Sales Amount"))) +
  geom_line(color = 3) +
  geom_point(color = 3) +
  theme(text = element_text(size = 18)) +
  scale_y_continuous(limits = c(0, 6000000), labels = comma) +
  geom_line(y=dataset$SalesAmountPred, color = 4) +
  geom_point(y=dataset$SalesAmountPred, color = 4) +
  theme(legend.position = "none") +
  labs(x = "Date",
       y = "Sales Amount")
```

See the result in Figure 9-11. The prediction (blue line) is considerably higher than the true values for the last two data points.

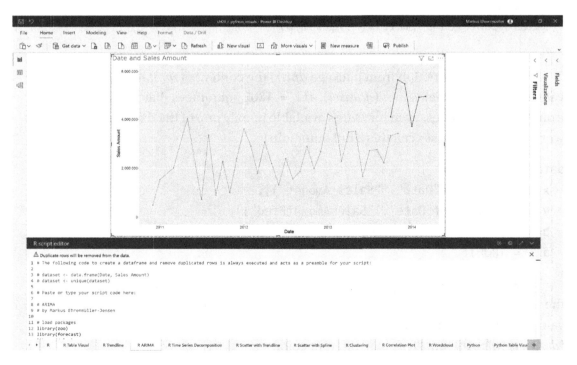

Figure 9-11. *R script visual showing an ARIMA forecast*

R Script Visual: Time-Series Decomposition

The *forecast* package makes it easy to decompose a time series into trend and seasonal parts.

```
# load package
library(forecast)
```

Data preparation involves the same steps as in the script of the previous section:

```
# prepare dataset
dataset$Date <- as.POSIXct(dataset$Date)
ts = ts(
    dataset["Sales Amount"],
    start = c(2011,1),
    frequency = 12)
```

Function decompose splits a time series into *observed* (actual values), *trend* (regression), *seasonal* (repeating patterns over the frequency), and *random* (part that cannot be explained by either trend or seasonal). The result set can easily be plotted with the base package's plot function:

```
# create plot
plot(decompose(ts))
```

In Figure 9-12, you can see how this looks. There is a clear upward trend over time (at least from the beginning of year 2012).

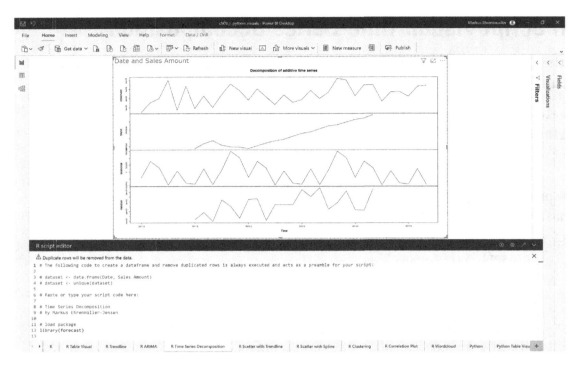

Figure 9-12. *R script visual showing a time-series decomposition*

R Script Visual: Scatter with Trendline

Here we will create a trendline on a scatter plot showing *Order Quantity* over *Unit Price*. We do not need to calculate the trendline separately, as package *ggplot2* is capable of doing this for us:

```
# load packages
library(ggplot2)
library(scales)
```

The plot is created exactly the same way as shown in section "R Script Visual Trendline," except the names of the columns changed (to *UnitPrice* and *OrderQuantity* from table *Reseller Sales*) and geom_line is not used (as it does not make very much sense to connect the data points in this case):

```
# create plot
ggplot(dataset,
       aes(x=UnitPrice,
           y=OrderQuantity)) +
  geom_point(color = 3) +
  theme(text = element_text(size = 18)) +
  theme(legend.position = "none") +
  labs(x = "Unit Price",
       y = "Order Quantity") +
  stat_smooth(method=lm, se=FALSE)
```

As expected, Figure 9-13 shows a negative correlation of unit price to order quantity: more expensive items are ordered in lower quantities.

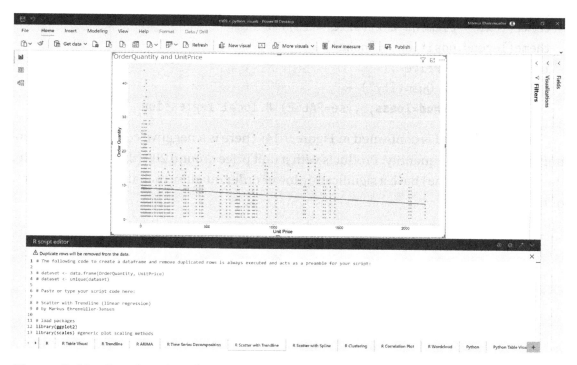

Figure 9-13. *R script visual: scatter with trendline*

R Script Visual: Scatter with Spline

Making a curved line (spline) instead of a straight line is a piece of cake in R with *ggplot2*. The only change to the script here, compared to the one in the previous section, is the last line: method=loess (for local regression), instead of lm (for linear model):

```
# load packages
library(ggplot2)
library(scales)

# create plot
ggplot(dataset,
       aes(x=UnitPrice,
           y=OrderQuantity)) +
  geom_point(color = 3) +
  theme(text = element_text(size = 18)) +
  theme(legend.position = "none") +
  labs(x = "Unit Price",
       y = "Order Quantity") +
  stat_smooth(method=loess, , se=FALSE) # local regression
```

The overall trend is confirmed in Figure 9-14: There is a negative correlation between unit price and order quantity. Products with a unit price around 200 (that's the area with the "valley" in the line) have a significantly lower order quantity than those with a lower or higher unit price.

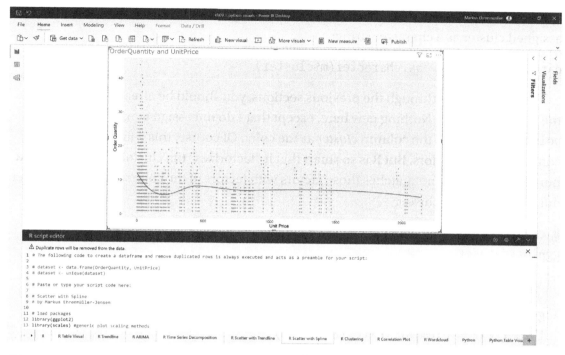

Figure 9-14. *R script visual: scatter with spline*

R Script Visual: Clustering

This example applies a k-means machine learning model to the data to cluster columns *OrderQuantity* over *UnitPrice*. The assigned cluster is then used to color the data points accordingly. First, we load the packages:

```
# load packages
library(ggplot2)
library(scales)
```

Then, we create a model *m* by calling the kmeans function from the base package. As parameters, we pass in the data frame *dataset* and that we want the model to find three clusters:

```
# create model
m <- kmeans(x = dataset, centers = 3)
```

Next, I create a new column *cluster* in the existing data frame *dataset* to show the assigned cluster as a character string:

```
dataset$cluster <- as.character(m$cluster)
```

If you have read through the previous sections, you should be already friends with the ggplot function. Nothing new here, except that I do not assign a fixed color in geom_point but rather use the column *cluster* as the color. Of course, this column does not contain names of colors. But R is so smart that it "factorizes" the clusters (a.k.a. assigns a numerical ID to it) and matches those factors with factors of the colors. Therefore, we get different colors per cluster.

```
# create plot
ggplot(dataset,
       aes(x=UnitPrice,
           y=OrderQuantity)) +
  geom_point(color = dataset$cluster) +
  theme(text = element_text(size = 18)) +
  theme(legend.position = "none") +
  labs(x = "Unit Price",
       y = "Order Quantity")
```

Figure 9-15 shows the scatter plot with three clusters highlighted in three different colors.

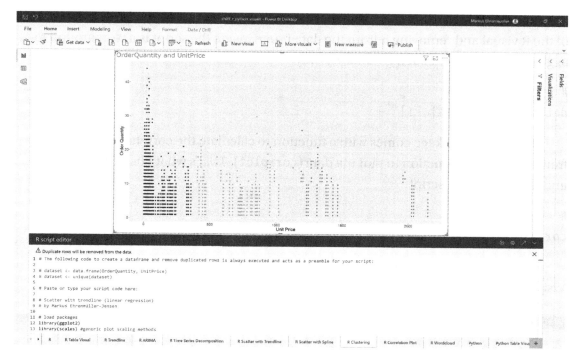

Figure 9-15. *R script visual: clustering*

R Script Visual: Correlation Plot

A correlation plot shows if there is a statistical relationship (cause-and-effect) between two variables and how strong the relationship is. Here, we are creating one for a bunch of measures (out of "table" *_Measures*): *Discount Amount, Freight, Order Quantity, Product Standard Cost, Sales Amount, Total Product Cost*, and *Unit Price AVG*. We only need a single package and two lines of code:

```
# load packages
library(corrplot)
```

If you only added the columns just mentioned, then the data frame *dataset* would consist of a single row only (with aggregated values of *Discount Amount*, etc.). Therefore, I added *Date* into the list (as first columns). Now I get one row per *Date* (which means one row per month in the available model). As I am not interested in any correlation between *Date* and the other columns (and to avoid a script error), I remove the first column (*Date*) from the *dataset* in the script. Adding *Date* as a field in the *Values* section of the R visual and removing it from the data frame inside the script is a workaround to aggregations on the needed granularity only.

```
# prepare dataset
dataset <- dataset[,-1]
```

The *corrplot* package comes with a function to calculate the correlation for a data frame (`cor`) and a function to plot the data (`corrplot`). That's what I used in the third and final line of the script:

```
# create plot
corrplot(cor(dataset))
```

Again, we get the same chart as in the sister section of Chapter 5 ("Adding Smart Visualizations"), as you can see in Figure 9-16.

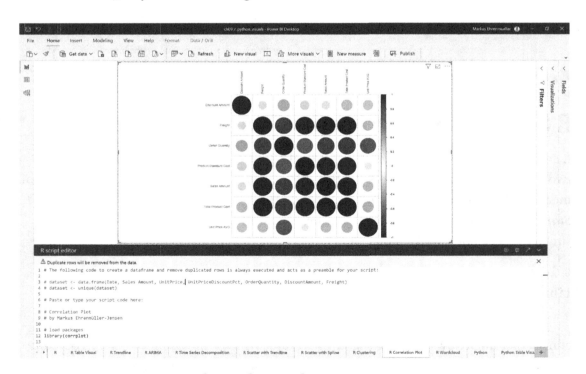

Figure 9-16. *R script visual: correlation plot*

R Script Visual: Word Cloud

Producing a word cloud is the easy part. Cleaning up the key phrases is the bigger challenge. In Chapter 5 ("Adding Smart Visualizations") I promised to show you a way to do so in later chapters. One version I will introduce now, while different ones will come in Chapter 11 ("Executing Machine Learning Models in the Azure Cloud").

I use two packages for this example. *tm* consists of functions for "text mining." *wordcloud* plots the actual word cloud.

```
# load packages
library(tm)
library(wordcloud)
```

VectorSource nested in VCorpus creates a list of words out of column *EnglishProductNameAndDescription* (from table *Product*). I created the column in the model by concatenating columns *EnglishProductName* and *EnglishProductDescription*. I did this because the description is sometimes empty and because the actual name is important as well.

```
# prepare data
vc <- VCorpus(VectorSource(dataset$EnglishProductNameAndDescription))
```

In the next steps, I use tm_map to transform all words into lowercase, remove numbers, remove stop words (stopwords() delivers a decent list for texts in English language), remove punctuations, and remove white space.

```
vc <- tm_map(vc, content_transformer(tolower))
vc <- tm_map(vc, removeNumbers)
vc <- tm_map(vc, removeWords, stopwords())
vc <- tm_map(vc, removePunctuation)
vc <- tm_map(vc, stripWhitespace)
```

The last cleaning step reduces words to their "stem." This means, for example, the words "provide," "provided," and "providing" are reduced to "provid" (without letter "e" at the end). No matter what version of the word is used in the text, they are all put into the same basket and counted together. While the single variants will only appear a few times and will get a low rank each, the stemmed version of the word might reach a prominent position, as all counts are summed up:

```
vc <- tm_map(vc, stemDocument)
```

Of course, all preceding steps are optional. If you do not like to have all the words in lowercase, if numbers are important in your text, or if stemming the words would confuse the report users, then just omit those steps. Having a script at hand gives you all the freedom. The final step produces the plot:

```
# create plot
wordcloud(vc, min.freq = 1, random.order = FALSE)
```

Figure 9-17 shows the word cloud for the cleaned product names and descriptions.

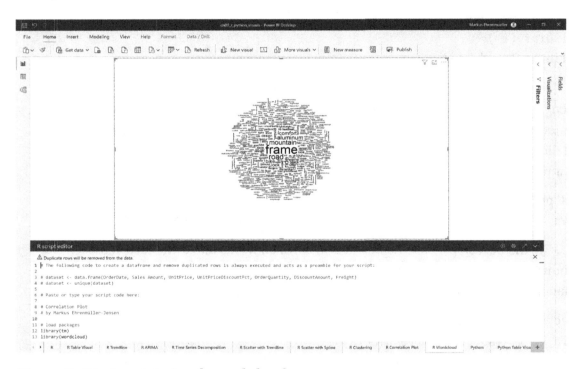

Figure 9-17. *R script visual: word cloud*

Python Script Visual: Table

For the rest of the chapter, we are back on track with Python script visuals. A simple script I sometimes use as the poor man's debugger is to plot the data (or intermediate transformations of the data frame) as a table. For developing and in-depth debugging, I would recommend using *Edit script in external IDE* instead. In the IDE, you can have multiple outputs, while a Python script visual allows for a single output only. We only need one package to load to plot data:

```
# import packages
import matplotlib.pyplot as plt
```

Then we create an empty plot, to which we "add" the table visual in a later step.

```
# create (empty) plot
fig = plt.figure(figsize=(15,6))
ax = fig.add_subplot(111)
ax.axis('off')
```

The table visual is then plotted for the values and columns of the dataset, located in the center of the empty plot. I set the font size explicitly to 12. Don't forget the plt.show at the end of the script.

```
# create table visual
MyTable = plt.table(
    cellText = dataset.values,
    colLabels = dataset.columns,
    loc = 'center')
MyTable.auto_set_font_size(False)
MyTable.set_fontsize(12)
plt.show()
```

The result is shown in Figure 9-18 and gives us a similar experience to when we created a table visual with the R script visual.

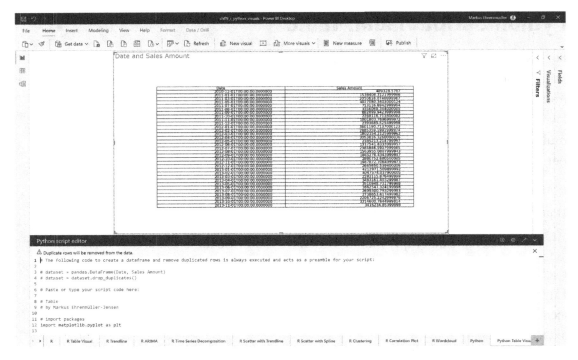

Figure 9-18. *Python script visual: table*

Python Script Visual: Trendline

The R script visuals in the previous sections did easily rebuild the exact examples from Chapter 5 ("Adding Smart Visualizations"), as the custom visualizations used in Chapter 5 are built upon R scripts (hidden in the code of the custom visualization). The following examples rebuild the same examples, but the result will look slightly different as they are built on Python (and the packages available there). You can of course tweak the Python scripts to come very close to the R script visuals. And you can tweak both the R and the Python script visuals to come close to Power BI's standard visuals. But this would be a topic for a book on its own. I tried to keep the scripts as simple as possible to achieve a working example.

Now let's get to the trendline. First, I load the necessary packages. *matplotlib.pyplot* we already have seen in the previous section and will be used in the remaining sections to actually plot the data (similar to what *ggplot2* is in R). *matplotlib.ticker* allows for business-oriented number formats (as opposed to scientific notation; similar to package *scale* in R). Data frames are part of R's base package. In Python, the functionality is covered in package *pandas*. *sklearn* delivers most of the machine learning algorithms I am using in the rest of the chapter. *sk* is the abbreviation for *SciKit*, which stands for SciPy Toolkits (https://www.scipy.org/scikits.html). *learn* is the hint that it contains machine learning algorithms.

```
# import packages
import matplotlib.pyplot as plt
from matplotlib.ticker import ScalarFormatter, FormatStrFormatter
import pandas as pd
from sklearn import linear_model
```

Then I convert column *Date* into the proper date-time format.

```
# prepare dataset
dataset.Date = pd.to_datetime(dataset.Date)
```

After that, I create a linear regression model *m* and train it with the actual data from *dataset*. Both columns (*Data* and *Sales Amount*) must be reshaped.

```
# create model
m = linear_model.LinearRegression()
m.fit(dataset.Date.values.reshape(-1, 1), dataset['Sales Amount'].values.
reshape(-1, 1))
```

Method `predict` applies the model *m* on the *Date* column. The result is then stored as a new column *prediction* in *dataset*.

```
# create prediction
m.pred = m.predict(dataset.Date.values.astype(float).reshape(-1, 1))
dataset['prediction'] = m.pred
```

The following code plots *Date* and *Sales Amount* in green as a line (style='-'):

```
# create plot
ax = dataset.plot(
    x='Date',
    y='Sales Amount',
    style='-',
    figsize=(15,8))
```

On top of this plot (*ax*), the predicted values are plotted in blue:

```
dataset.plot(
    x='Date',
    y='prediction',
    color='blue',
    ax=ax)
```

The rest of the script sets labels for the x- and y-axis and applies a common number format to the y-axis.

```
ax.set_xlabel('Date')
ax.set_ylabel('Sales Amount')
ax.get_yaxis().set_major_formatter(matplotlib.ticker.FuncFormatter(lambda
x, p: format(int(x), ',')))
plt.show()
```

Sales Amount and a *prediction* in the form of a trendline are plotted in Figure 9-19.

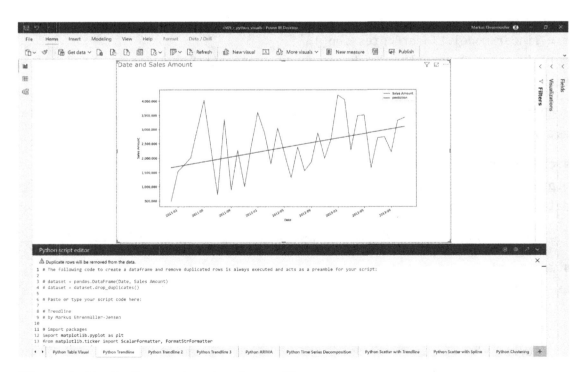

Figure 9-19. *Python script visual: trendline*

Python Script Visual: ARIMA

In this section, we will create a prediction with an Auto-regressive Integrated Moving Average (ARIMA). This algorithm is available in package *statsmodels.api*. The rest of the packages we already know from the previous sections of this chapter.

```
# import packages
import matplotlib.pyplot as plt
from matplotlib.ticker import ScalarFormatter, FormatStrFormatter
import pandas as pd
import statsmodels.api as sm
```

For the model, it is not enough to just convert *Date* into a date-time data type. It is necessary to set an index based on the resampled *Date* column. MS stands for month start, which makes sure that all the *Sales Amount* is moved to the beginning of the month (as the sales are scattered on different days of the month, but not on enough days per month to make a model based on daily data feasible).

```
# prepare dataset
dataset.Date = pd.to_datetime(dataset.Date)
dataset = dataset.set_index('Date').resample('MS').pad()
```

y is the variable used to feed the model. All the sales scattered within a month are aggregated onto the first of the month (again, MS stands for month start).

```
y = dataset['Sales Amount'].resample('MS').sum()
```

ARIMA's parameter for *order* (p, d, q) and *seasonal_order* (P, D, Q) are set here. The seasonality is set to twelve periods, as we have monthly data.

```
# create model
m = sm.tsa.statespace.SARIMAX(
        y,
        order=(0, 1, 0),
        seasonal_order=(0, 1, 0, 12),
        )
```

A prediction (forecast) for the next six months is created in the next step:

```
# create forecast
m.fit = m.fit().get_forecast(steps=6)
```

First, the actual sales amount is plotted in green:

```
# create plot
ax = y.plot(label='Sales Amount', color='green')
```

Then, the predicted values are added to the existing plot:

```
m.fit.predicted_mean.plot(ax=ax, label='prediction')
```

The following lines of code are the same as in the previous section and make sure that the labels for the axis, the number format, and the legend are plotted in the right way:

```
ax.set_xlabel('Date')
ax.set_ylabel('Sales Amount')
ax.get_yaxis().set_major_formatter(matplotlib.ticker.FuncFormatter(lambda
x, p: format(int(x), ',')))
plt.legend()
plt.show()
```

The actual values and the forecast as predicted by ARIMA are plotted in Figure 9-20.

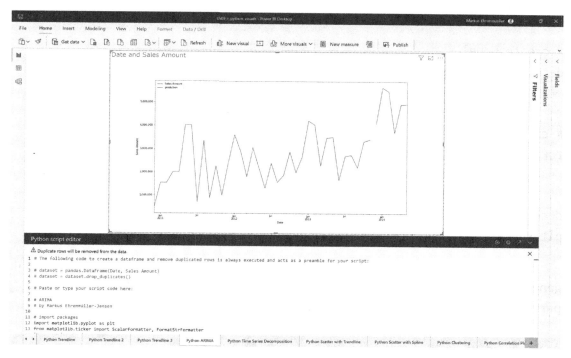

Figure 9-20. *Python script visual: ARIMA*

Python Script Visual: Time-Series Decomposition

In Python, we can see the components of what makes a time series (and what is used by the model to predict and forecast) in the form of a plot. I imported *rcParams* to be able to set the size proportions of the figure.

```
# import packages
import matplotlib.pyplot as plt
import pandas as pd
import statsmodels.api as sm
from pylab import rcParams
```

The next lines to prepare the dataset are equal to what we saw in the previous section:

```
# prepare dataset
dataset.Date = pd.to_datetime(dataset.Date)
dataset = dataset.set_index('Date').resample('MS').pad()
y = dataset['Sales Amount'].resample('MS').sum()
```

The model is created via function `sm.tsa.seasonal_decompose`:

```
# create model
decomposition = sm.tsa.seasonal_decompose(y, model='additive')
```

First, I set the size of the figure to fill the available space as well as possible:

```
# create plot
rcParams['figure.figsize'] = 12, 6
```

Then, I plot the model:

```
decomposition.plot()
plt.show()
```

The plot in Figure 9-21 shows the discovered parts of the sales amount time series. In general, this shows a positive *Trend* over time and repeating *Seasonal* patterns. And the parts not explained by *Trend* and *Seasonal* are *Resid*, short for residual.

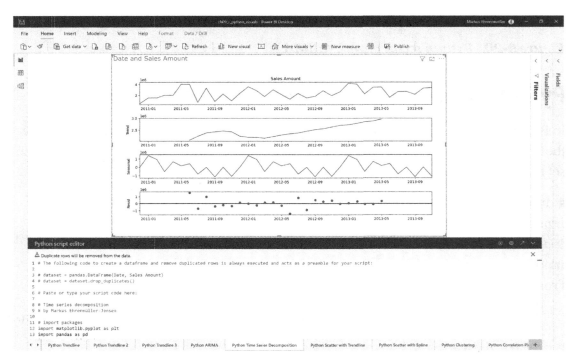

Figure 9-21. *Python script visual: time-series decomposition*

Python Script Visual: Scatter with Trendline

Here comes the Python script visual to show *Reseller Sales' OrderQuantity* over *UnitPrice* with a straight trendline. The whole script is very similar to the one for the trendline earlier. *matplotlib* contains all we need to plot the data. And we use *sklearn* for calculating the values for the trendline.

```
# import packages
import matplotlib.pyplot as plt
from sklearn import linear_model
```

The values for the trendline are calculated as a simple linear regression model, which is trained with *UnitPrice* and *OrderQuantity*:

```
# create model
m = linear_model.LinearRegression()
m.fit(dataset['UnitPrice'].values.reshape(-1, 1), dataset['OrderQuantity'].
values.reshape(-1, 1))
```

After that, the trained model is applied to the actual data and stored as a new column in the existing data frame *dataset*:

```
# create prediction
m.pred = m.predict(dataset.UnitPrice.values.astype(float).reshape(-1, 1))
dataset['prediction'] = m.pred
```

Two lines are than plotted: the actual data as green dots and the predicted trend as a line:

```
# create plot
ax = dataset.plot(
    x='UnitPrice',
    y='OrderQuantity',
    color='green',
    style='.',
    figsize=(13,6))
```

```
dataset.plot(
    x='UnitPrice',
    y='prediction',
    ax=ax
    )
```

Both axes are labeled with the right text, and the legend gets removed:

```
ax.set_xlabel('Unit Price')
ax.set_ylabel('Order Quantity')
ax.get_legend().remove()
plt.show()
```

You can enjoy the result of the script in Figure 9-22, which shows an indirect proportional relationship between unit price and order quantity.

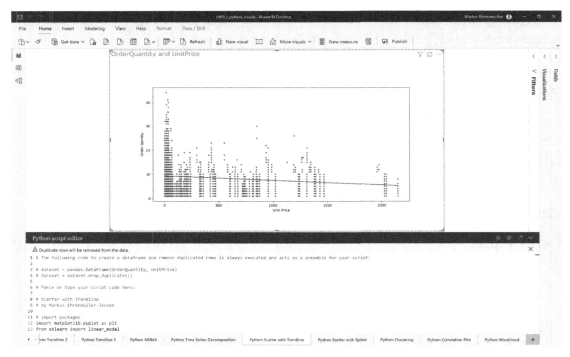

Figure 9-22. *Python script visual: scatter with trendline*

Python Script Visual: Scatter with Spline

Instead of a straight trendline, we can use a different algorithm to generate a curved line. Depending on your use case, this might give better information about the relationship between two variables. For this purpose, I added two additional packages from *sklearn*:

```
# import packages
import matplotlib.pyplot as plt
from sklearn.linear_model import LinearRegression
from sklearn.preprocessing import PolynomialFeatures
from sklearn.pipeline import make_pipeline
```

If you do not sort the data frame, the resulting spline will go back and forth on the x-axis—not what is useful in our case. That's why I sort the rows of the data frame as follows:

```
# prepare dataset
dataset = dataset.sort_values(by=['UnitPrice'])
```

The model is still a linear regression, but a polynomial of the fifth degree.

```
# create model
m = make_pipeline(PolynomialFeatures(degree = 5), LinearRegression())
m.fit(dataset['UnitPrice'].values.reshape(-1, 1), dataset['OrderQuantity'].
values.reshape(-1, 1))
```

Creating a prediction with this model uses the same code as with the simple linear regression model:

```
# create prediction
m.pred = m.predict(dataset.UnitPrice.values.astype(float).reshape(-1, 1))
dataset['prediction'] = m.pred
```

There isn't really anything to learn here. The predicted values are different, but the code to plot it is the very same as in the previous section:

```
# create plot
ax = dataset.plot(
    x='UnitPrice',
    y='OrderQuantity',
```

```
    color='green',
    style='.',
    figsize=(13,6))
dataset.plot(
    x='UnitPrice',
    y='prediction',
    ax = ax
    )
ax.set_xlabel('Unit Price')
ax.set_ylabel('Order Quantity')
ax.get_legend().remove()
plt.show()
```

The spline is applied in Figure 9-23.

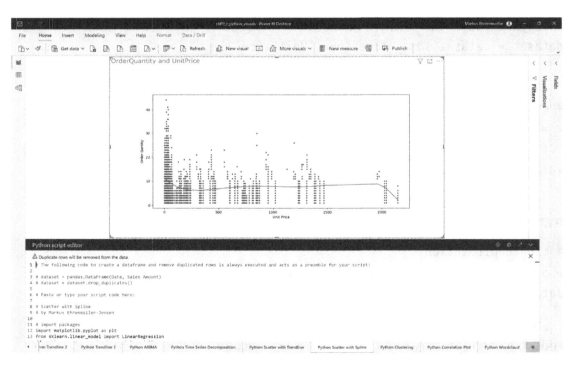

Figure 9-23. *Python script visual: scatter with spline*

Python Script Visual: Clustering

For clustering, I chose a k-means algorithm from the *sklearn* package:

```
# import packages
import matplotlib.pyplot as plt
from sklearn.cluster import KMeans
```

I ask function KMeans to find three clusters in *dataset*. What the right or wrong number of clusters would be is totally up to yourself and the use case.

```
# create model
kmeans = KMeans(n_clusters=3).fit(dataset)
```

The outcome is then plotted as a scatter plot, with the clusters assigned to parameter *c*, which controls the color. I set parameter *alpha* to 0.3 to make the data points translucent, which allows one to see where overlapping (in dense areas of the chart) happens.

```
# create plot
fig = plt.figure(figsize=(13,6))
plt.scatter(
    dataset['UnitPrice'],
    dataset['OrderQuantity'],
    c = kmeans.labels_.astype(float),
    alpha = 0.3)
plt.show()
```

The k-means implementation used in this example is the same as in the example previously seen in R. Figure 9-24 shows the same three clusters.

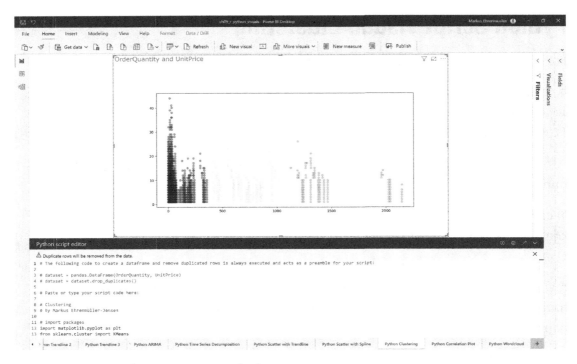

Figure 9-24. *Python script visual: clustering*

Python Script Visual: Correlation Plot

The correlation plot in the following Python script looks a bit different from the layout in R, but shows the same information and the same correlation as previously seen with the R script visual. I built it upon *Discount Amount, Freight, Order Quantity, Product Standard Cost, Sales Amount, Total Product Cost,* and *Unit Price AVG*. Column *Date* is added to get the granularity from aggregated totals down to aggregations per date (a.k.a. month). We meet a new package here: *seaborn.*

```
# import packages
import matplotlib.pyplot as plt
import pandas as pd
import seaborn as sns
```

Convert column *Date* to a date-time, just as we have seen multiple times in the previous sections:

```
# prepare dataset
dataset.Date = pd.to_datetime(dataset.Date)
```

Here, I apply *seaborn*'s corr function to calculate the correlations between all columns of data frame *dataset*:

```
# create model
corr = dataset.corr()
```

The plot is then done as a heatmap. Squares have a background color that depends on the correlation coefficient. *vmin* and *vmax* are set to the theoretical minimum and maximum of a correlation coefficient (-1 and 1). If you omit both parameters, the color scale will only cover the range of the existing values. I set the center to 0 to make sure that a correlation coefficient of 0 is plotted in white. sns.diverging_palette with parameters 10 and 240 sets the color map (cmap) range between a saturated blue and a saturated red; we get n=200 different shades in between. I opted to format the table as a square.

```
# create plot
ax = sns.heatmap(
    corr,
    vmin=-1,
    vmax=1,
    cmap=sns.diverging_palette(0, 250, n=200),
    square=True
)
```

ax.set_xticklabels makes sure to put the names on the x axis in a readable format.

```
ax.set_xticklabels(
    ax.get_xticklabels(),
    rotation=45,
    horizontalalignment='right'
)
plt.show()
```

You see the result of the script in Figure 9-25. The layout is a bit different than the R version of it. The information is completely the same, though.

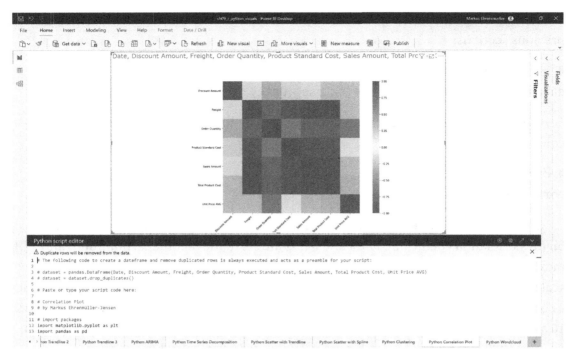

Figure 9-25. *Python script visual: correlation plot*

Python Script Visual: Word Cloud

The final script in this chapter generates a word cloud from *Product*'s
EnglishProductNameAndDescription. I use a package of the same name:

```
# import packages
import matplotlib.pyplot as plt
import pandas as pd
from wordcloud import WordCloud, STOPWORDS
```

The actual stop words are imported from the *wordcloud* package:

```
# prepare dataset
stopwords = set(STOPWORDS)
```

I then initialize a variable *KeyPhrases* and fill it with the token extracted by looping over the content of *EnglishProductNameAndDescription* of data frame *dataset*. Function split separates a string by white spaces. The only transformation I apply is to convert the strings to lowercase.

```
KeyPhrases = ''
for val in dataset['EnglishProductNameAndDescription']:
    val = str(val)
    tokens = val.split()
    for i in range(len(tokens)):
        tokens[i] = tokens[i].lower()
    KeyPhrases += " ".join(tokens)+" "
```

Function WordCloud creates an object containing the information needed for the word cloud (key phrases and word counts). The parameters are rather self-explanatory. imshow plots this object.

```
# create plot
wordcloud = WordCloud(
    width = 1500,
    height = 1000,
    background_color ='white',
    stopwords = stopwords,
    min_font_size = 10).generate(KeyPhrases)
plt.imshow(wordcloud, interpolation='bilinear')
plt.axis("off")
plt.show()
```

Figure 9-26 shows the result of the script: a rectangle and colorful word cloud.

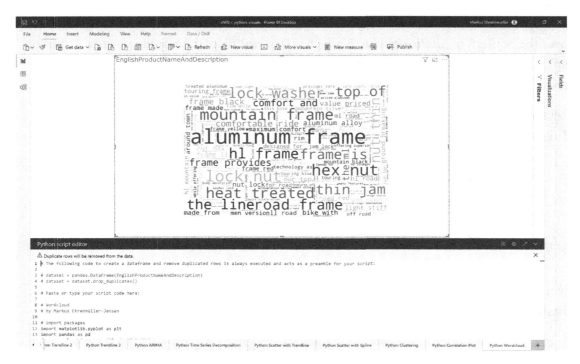

Figure 9-26. *Python script visual: word cloud*

Power BI Service and Power BI Report Server

R and Python visuals are supported in Power BI Service as well. If you intend to publish reports to the service, please check carefully whether the packages you are planning to use are available in the service—and, if so, which version is available. The supported packages in the service are out of your control. Always use the supported versions of the packages on your computer to avoid surprises after publishing.

This link provides you with a list of R packages supported in Power BI Service: `https://docs.microsoft.com/en-us/power-bi/connect-data/service-r-packages-support`. And this link is about Python packages in Power BI Service: `https://powerbi.microsoft.com/en-us/blog/python-visualizations-in-power-bi-service/`.

You can use R and Python visuals in Power BI Desktop for Report Server, but the visuals will not work when published to Power BI Report Server. At time of writing, this feature was marked as "not planned" on `ideas.powerbi.com`. Find a list of differences between Power BI Service and Power BI Report Server here: `https://docs.microsoft.com/en-us/power-bi/report-server/compare-report-server-service`.

Key Takeaways

R and Python script visuals inside Power BI Desktop open up a whole new world, as follows:

- R visualization: You can run a fully supported R script inside a visualization in Power BI, which generates a plot. R and all packages must be installed on your computer.

- Python visualization: You can run a fully supported Python script inside a visualization in Power BI, which generates a plot. Python and all packages must be installed on your computer.

- *Edit script in external IDE* is a cool feature to connect Power BI with an R and/or Python editor of your choice. Not only is the script inside the visual opened in the external IDE, but the current data is exported as well, and lines of code are automatically added at the beginning of the script to import the data.

- R, Python, and the most common packages are available in Power BI Service.

- Power BI Report Server does not support R and/or Python visuals.

Not enough from R and Python? Learn how to use R and Python in Power Query in the next chapter.

## CHAPTER 10

Transforming Data with R and Python

Power BI, R, and Python have in common that they are heavily used to ingest, transform, and prepare data. All three come with very helpful ways of achieving such tasks. You can use Power BI's core functionalities to transform data either via the graphical interface or by writing M code, as you learned in Chapter 8 ("Creating Columns by Example"). If you hit the limitations of Power Query, you can use a transformation written in R or Python, or even reuse one you already have, inside Power Query.

R and Python

Before you continue reading this chapter, make sure you understand the explanations given in the first sections of Chapter 9 ("Executing R and Python Visualizations"). All hints on what to install and what options to set before you can start with the examples are valid for this chapter as well.

R and Python can be used not only inside a visual, but in Power Query as well. This is what we will concentrate on in this chapter. We will start with R. You can find explanations and examples for Python in the second half of this chapter.

Whether you use scripts in R or in Python, all will more or less contain the following steps (single steps might be omitted):

- Load packages/libraries

- Prepare dataset

- Create model

- Create output

© Markus Ehrenmueller-Jensen 2020
M. Ehrenmueller-Jensen, *Self-Service AI with Power BI Desktop*, https://doi.org/10.1007/978-1-4842-6231-3_10

All data from the previous step in Power Query (if you use a script transformation) is exposed as a data frame *dataset* (similar to a script visual, as described in the previous chapter).

The last step is mandatory to return the result to the next step (in both a script source or a script transformation). In an R script transformation, the script must create a new data frame—otherwise, Power Query cannot receive the result. In Python, everything can be added to data frame *dataset*.

Load Data with R

In Figure 10-1, you can see the content of step *Source* of query *R Source*. It is a simple R script with a single line of code calling function `read.csv` to load the content of the given path and file it into a data frame of name *dataset*. We can also see that *dataset* is exposed in the form of a *Table* in Power Query, which is then expanded into its columns by the next step (*Navigation*).

Here is the code from the R script. You need to change the path (marked in **bold**) accordingly if you want to try it out on your own. Make sure to either have forward slashes (/) or double backslashes (\\) to separate the elements in the path. You can change the path either in the formula bar via the gear icon at step *Source* or in the Advanced Editor:

```
# load data
dataset = read.csv('C:/Users/mehre/OneDrive - Markus Ehrenmueller(1)/Talks/
Power BI AI/Book/ch10_r_python_power_query_demo/Date&SalesAmount.csv',
check.names = FALSE, encoding = "UTF-8", blank.lines.skip = FALSE);
```

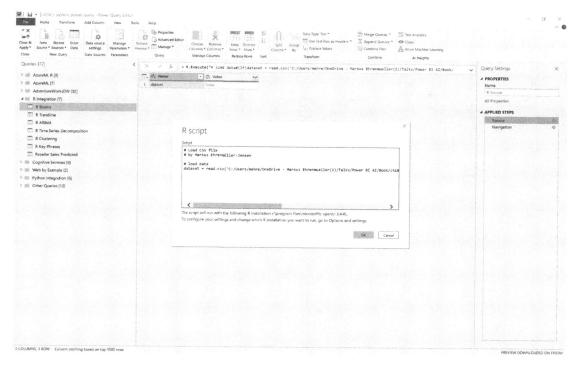

Figure 10-1. *R script source*

Note   This example, loading data from a CSV file via an R script, is for educational purposes to demonstrate a simple script. I would prefer to achieve the task with native Power Query functionalities instead (e.g., *Get Data* ➤ *Text/CSV* in the ribbon).

If you want to load data to Power BI with an R script, use the following steps:

- Select *Home* ➤ *Transform Data* in Power BI's ribbon to start Power Query (if you have not already done so).

- Select *Home* ➤ *New Source* in Power Query. A long list of possible data sources is shown in a flyout window. Use the search field or the scroll bar to find and select *R script* (Figure 10-2).

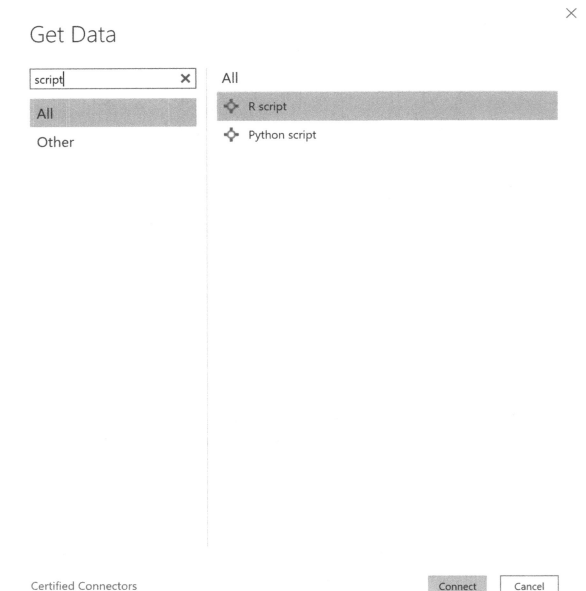

Figure 10-2. *Get data from R or Python scripts*

- Enter (or paste) the script of your choice. Unfortunately, this editor is not much more than a textbox (i.e., no syntax highlighting), and we cannot start an external IDE here (Figure 10-3).

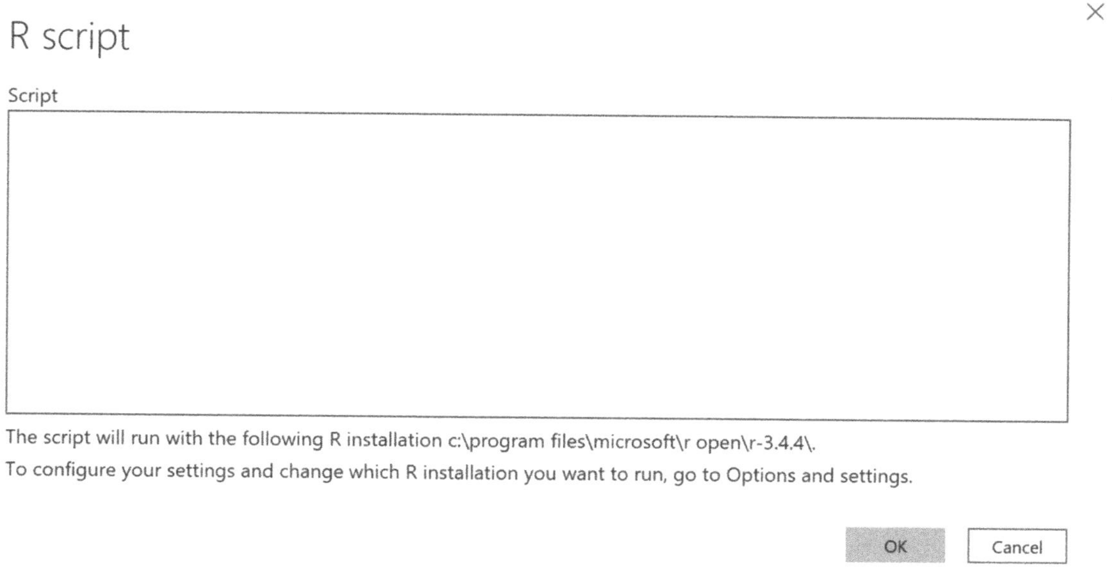

R script

×

Script

The script will run with the following R installation c:\program files\microsoft\r open\r-3.4.4\.

To configure your settings and change which R installation you want to run, go to Options and settings.

OK   Cancel

Figure 10-3. *R script editor*

- The result of this step is a list of all data frames created by the script (already shown in Figure 10-1). *Name* shows the name of the data frame. *Value* shows *Table*. You can click on the cell containing *Table* (not on the text itself) to get a preview of the content. Click on the text *Table* itself to expand the whole data frame. You get all rows and columns contained in the data frame and can further transform the content if you like. This is shown in Figure 10-4.

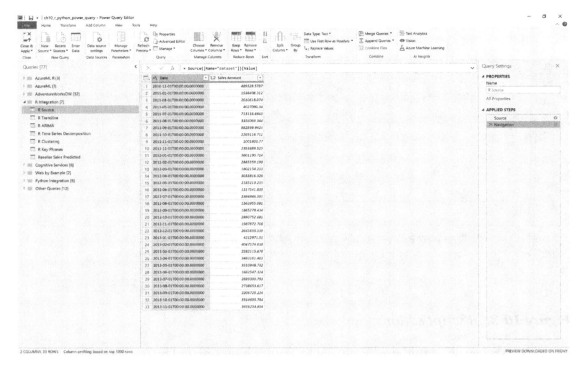

Figure 10-4. *R script result*

- Select *Close & Apply* in the ribbon to close the Power Query window and load the new data into Power BI's data model. A new table is shown in the field list with all selected columns of the data frame.

- Make sure to set the correct relationships in Power BI's model view. I created a *one-to-many* relationship between column *Date* in table *Date* and column *Date* of the new table.

Transform Data with R

In Figure 10-5, you can see the content of the step *Run R script* of query *R Trendline*. The result set from the previous step is exposed to this script as data frame *dataset*. The R script later creates data frame *output*, which contains columns *Date*, *Sales Amount*, and *prediction*. The last column contains the values for a trendline, generated as a linear regression. The data frame is exposed in the form of a *Table* to Power Query, which is then expanded into its columns by the next step (*Navigation*). The content of this script, and more examples, will be discussed in detail in the following sections.

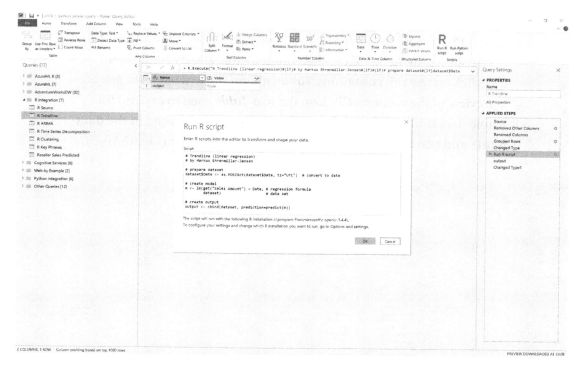

Figure 10-5. *R script transformation*

If you want to transform data in Power Query with an R script, use the following steps:

- Select *Home* ➤ *Transform Data* in Power BI's ribbon to start Power Query (if you have not already done so).

- Create a new query or find an existing one. Add a step by selecting *Transform* ➤ *Run R script* in Power Query (as shown in Figure 10-5).

- Enter (or paste) the script of your choice. Unfortunately, we cannot start an external IDE in this step. The content from the previous step is exposed to the script as data frame *dataset*. You must create a data frame of a different name (I usually choose *output*) to make data available for the subsequent step in the Power Query (as shown in Figure 10-5).

- The result is a list of all data frames created by the script (the input data frame *dataset* is not listed; that's why we must create a new one). *Name* shows the name of the data frame. *Value* shows *Table*. You can click on the cell containing *Table* (not on the text itself) to get a preview of the content. Click on the text *Table* itself to expand the whole data frame. You get all rows and columns contained in the data frame and can further transform the content if you like (Figure 10-6).

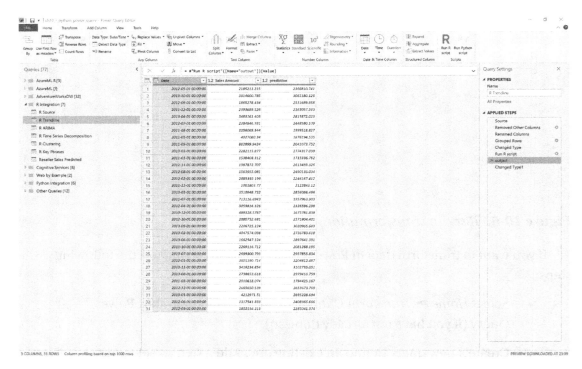

Figure 10-6. *R script transformation result*

- Select *Close & Apply* in the ribbon to close the Power Query window and load the new data into Power BI's data model. A new table is shown in the field list with all selected columns of the data frame.

- Make sure to set the correct relationships in the Power BI's model view. I created a *one-to-many* relationship between column *Date* in table *Date* and column *Date* of the new table.

Trendline

The R script in Power Query *R Trendline* (in step *Run R Script*) only has one line of code in common with the script in the "Trendline" section in Chapter 9. There, the trendline was generated inside ggplot, which we can't use here, as we are not going to plot data, but rather create the *output* data frame. The script uses the functionality of the base package only—no need to load packages. But we do have to convert column *Date* as we have done in almost all R scripts in this book:

```
# prepare dataset
dataset$Date <- as.POSIXct(dataset$Date, tz="UTC")
```

I invoke function lm of R's base package to create the model *m*. It creates a linear regression model. The first parameter is the formula. The tilde symbol (~) means "by." We are looking for a model that explains *Sales Amount* by *Date*. The second parameter is the data frame containing the data to train the model.

```
# create model
m <- lm(get("Sales Amount") ~ Date, dataset)
```

The *output* data frame is created then via function cbind, which binds new columns to an existing data frame. The result of function prediction (with the generated model *m* as a parameter) is added as a new column (with name *prediction*) to *dataset* and stored as *output*.

```
# create output
output <- cbind(dataset, prediction=predict(m))
```

We already have seen the result of the script, in Figure 10-6.

Time-Series Decomposition

Query *R Time Series Decomposition* contains the following script. It loads package *forecast* to create the time-series decomposition (as in the sister section in Chapter 9). I later use function fortify from package *ggfortify*.

```
# load packages
library(forecast)
library(ggfortify)
```

Preparation is done exactly as in Chapter 9. Column *Date*'s data type is converted, and function `ts` decomposes the time series.

```
# prepare dataset
dataset$Date <- as.POSIXct(dataset$Date)
ts = ts(
    dataset["Sales Amount"],
    start = c(2011,1),
    frequency = 12)
```

Function `fortify` adds the information from *ts* as new columns to existing data frame *dataset*. The result is stored in data frame *output*.

```
# create output
output <- fortify(decompose(ts), dataset)
```

The result in Figure 10-7 shows three new columns: *trend, seasonal,* and *remainder.*

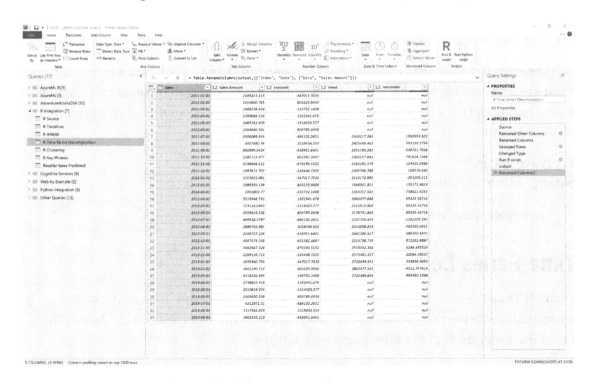

Figure 10-7. *R script has added three new columns.*

Clustering

Clustering is done again through a k-means algorithm (exactly as in Chapter 9). You can see the full script in query *R Clustering*. No special package is needed, as R's base package comes with the function kmeans out of the box. The model is created over *dataset* and asked to assign three clusters, as follows:

```
# create model
m <- kmeans(x = dataset, centers = 3)
```

The cluster is then stored as a new column in data frame *dataset*:

```
dataset$cluster <- as.character(m$cluster)
```

Everything is then copied into data frame *output*:

```
# create output
output <- dataset
```

Every existing pair of *OrderQuantity* and *UnitPrice* is assigned to a cluster, shown in Figure 10-8.

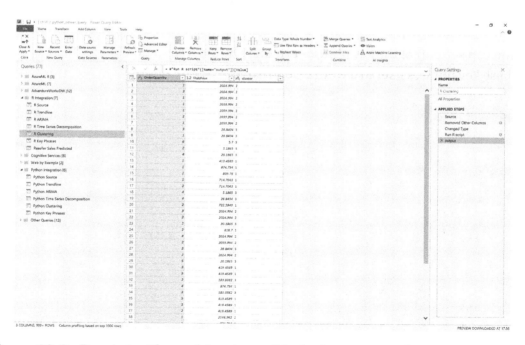

Figure 10-8. *R script with combinations of OrderQuantity and UnitPrice and their clusters*

Key Phrases

Extracting key phrases is not much different than what we did inside the R script visual in Chapter 9. Instead of plotting a word cloud, we return a data frame named *output* in this example. (We will plot the key phrases in a later section in this chapter.)

```
# load package
library(tm)

# prepare data
vc <- VCorpus(VectorSource(dataset$EnglishProductNameAndDescription))
vc <- tm_map(vc, content_transformer(tolower))
vc <- tm_map(vc, removeNumbers)
vc <- tm_map(vc, removeWords, stopwords())
vc <- tm_map(vc, removePunctuation)
vc <- tm_map(vc, stripWhitespace)
vc <- tm_map(vc, stemDocument)

# create output
output <- data.frame(text = sapply(corpus_clean, paste, collapse = " "),
stringsAsFactors = FALSE)
```

You can see part of the key phrases extracted by the R script in Figure 10-9. We will use these key phrases later in a word cloud visual.

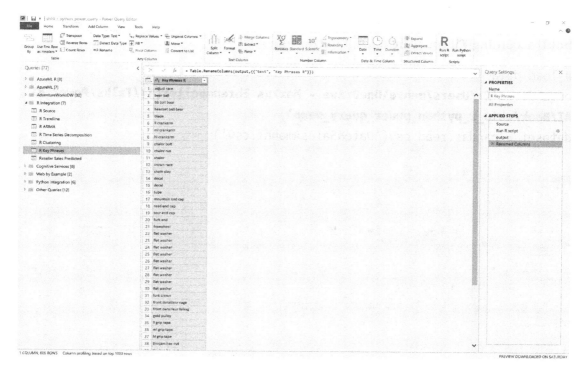

Figure 10-9. *R script's output of key phrases*

Load Data with Python

Now we will change the topic from R to Python. In Figure 10-10, you can see the content of step *Source* of query *Python Source*. It is a simple Python script with two lines of code calling function `os.chdir` to change the folder to the given path and `read.csv` to load the content of the given file into a data frame of name *dataset*. We can also see that *dataset* is exposed in the form of a *Table* in Power Query, which is then expanded into its columns by the next step (*Navigation*).

Here is the code from the Python script. Make sure to change the path (marked in **bold**) accordingly if you want to try it out on your own:

```
# load data
os.chdir(u'C:/Users/mehre\OneDrive - Markus Ehrenmueller(1)/Talks/Power BI
AI/Book/ch10_r_python_power_query_demo')
dataset = pandas.read_csv('Date&SalesAmount.csv')
```

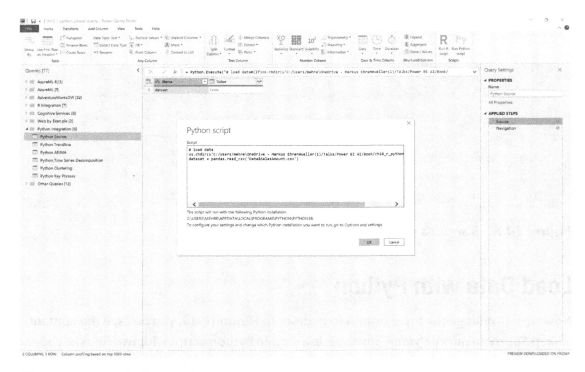

Figure 10-10. *Python script source*

Note   This example, loading data from a CSV file via a Python script, is for educational purposes to demonstrate a simple script. I would prefer to achieve the task with native Power Query functionalities instead (e.g., *Get Data* ➤ *Text/CSV* in the ribbon).

If you want to load data to Power BI with a Python script, use the following steps:

- Select *Home* ➤ *Transform Data* in Power BI's ribbon to start Power Query (if you have not already done so).

- Select *Home* ➤ *New Source* in Power Query. A long list of possible data sources is shown in a flyout window. Use the search field or the scroll bar to find and select *Python script* (Figure 10-11).

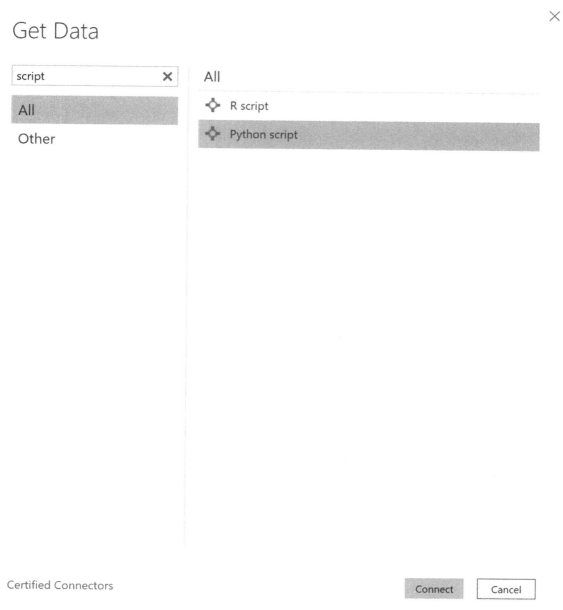

Figure 10-11. *Get data from R or Python script*

- Enter (or paste) the script of your choice. Unfortunately, this editor is not much more than a textbox (i.e., no syntax highlighting), and we cannot start an external IDE here (Figure 10-12).

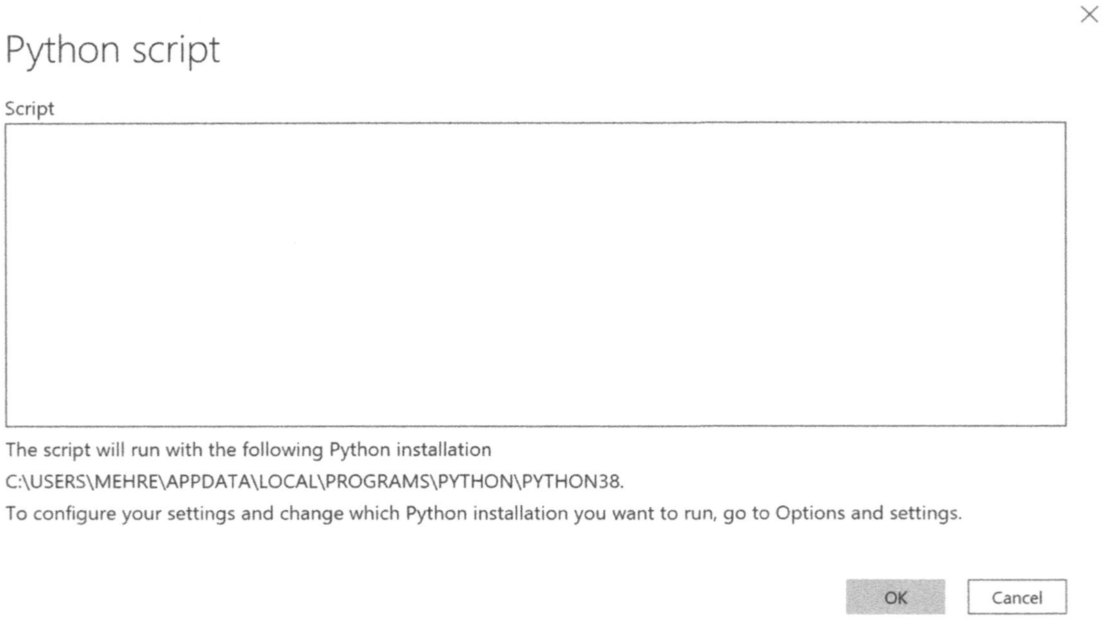

×

Python script

Script

The script will run with the following Python installation
C:\USERS\MEHRE\APPDATA\LOCAL\PROGRAMS\PYTHON\PYTHON38.
To configure your settings and change which Python installation you want to run, go to Options and settings.

OK   Cancel

Figure 10-12. *Python script editor*

- The result is a list of all data frames created by the script (already shown in Figure 10-10). *Name* shows the name of the data frame. *Value* shows *Table*. You can click on the cell containing *Table* (not on the text itself) to get a preview of the content. Click on the text *Table* itself to expand the whole data frame. You get all rows and columns contained in the data frame and can further transform the content if you like. This is shown in Figure 10-13.

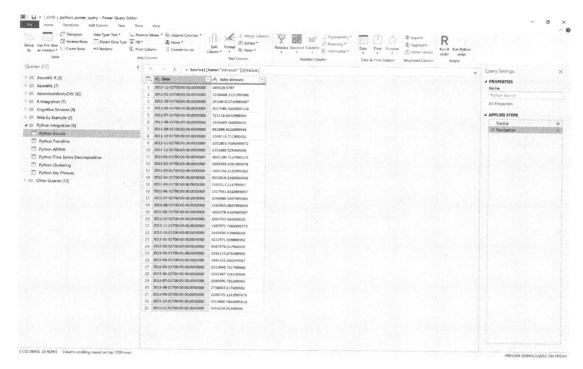

Figure 10-13. *Python script result*

- Select *Close & Apply* in the ribbon to close the Power Query window and load the new data into Power BI's data model. A new table is shown in the field list with all selected columns of the data frame.

- Make sure to set the correct relationships in Power BI's model view. I created a *one-to-many* relationship between column *Date* in table *Date* and column *Date* of the new table.

Transform Data with Python

In Figure 10-14, you can see the content of step *Run Python script* of query *Python Trendline*. The result from the previous step is exposed to this script as data frame *dataset*. The Python script later creates data frame *output*, which contains columns *Date*, *Sales Amount*, and *prediction*. The last column contains the values for a trendline, generated as a linear regression. The data frame is exposed in the form of a *Table* to Power Query, which is then expanded into its columns by the next step (*Navigation*). The content of the script, and more examples, will be discussed in upcoming sections.

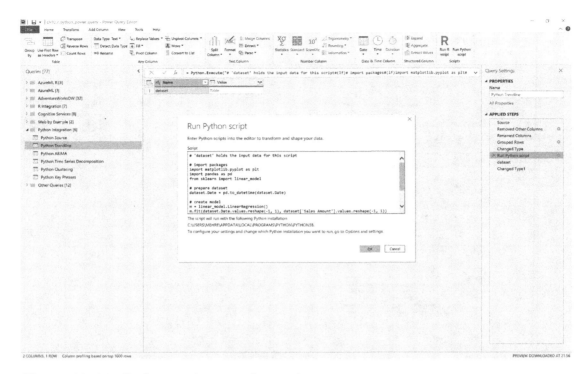

Figure 10-14. *Python script transformation*

If you want to transform data in Power Query with a Python script, use the following steps:

- Select *Home* ➤ *Transform Data* in Power BI's ribbon to start Power Query (if you have not already done so).

- Create a new query or find an existing one. Add a step by selecting *Transform* ➤ *Run Python script* in Power Query (as shown in Figure 10-14).

- Enter (or paste) the script of your choice. Unfortunately, we cannot start an external IDE in this step. The content from the previous step is exposed to the script as data frame *dataset*. You must create a data frame of a different name (I usually choose *output*) to make data available for the subsequent step in the Power Query (as shown in Figure 10-14).

- The result is a list of all data frames created by the script (including the input data frame *dataset*; as opposed to R we do not need to create a new one). *Name* shows the name of the data frame. *Value* shows *Table*. You can click on the cell containing *Table* (not on the text itself) to get a preview of the content. Click on the text *Table* itself to expand the whole data frame. You get all rows and columns contained in the data frame and can further transform the content if you like (Figure 10-15).

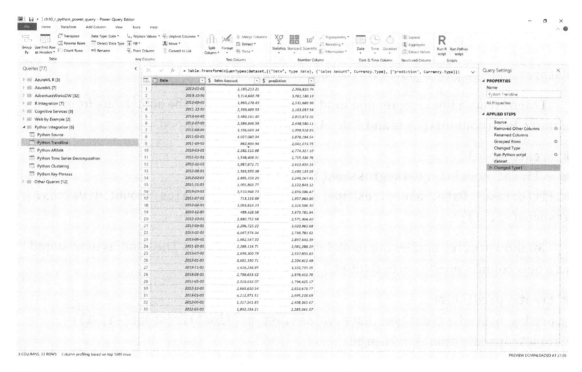

Figure 10-15. *Python script transformation result*

- Select *Close & Apply* in the ribbon to close the Power Query window and load the new data into Power BI's data model. A new table is shown in the field list with all selected columns of the data frame.

- Make sure to set the correct relationships in the Power BI's model view. I created a *one-to-many* relationship between column *Date* in table *Date* and column *Date* of the new table.

Trendline

The Python script in Power Query *Python Trendline* (in step *Run Python Script*) is the exact same as the script in the "Trendline" section in Chapter 9, except for the last lines, where the plot is created:

```
# import packages
import pandas as pd
from sklearn import linear_model
```

Then, I convert column *Date* into the proper datetime format.

```
# prepare dataset
dataset.Date = pd.to_datetime(dataset.Date)
```

I then create a linear regression model *m* and train it with the actual data from *dataset*. Both columns (*Date* and *Sales Amount*) must be reshaped.

```
# create model
m = linear_model.LinearRegression()
m.fit(dataset.Date.values.reshape(-1, 1), dataset['Sales Amount'].values.
reshape(-1, 1))
```

Method predict applies the model *m* on the *Date* column. The result is then stored as a new column *prediction* in *dataset*.

```
# create prediction
m.pred = m.predict(dataset.Date.values.astype(float).reshape(-1, 1))
dataset['prediction'] = m.pred
```

We already saw the result of this script in Figure 10-15.

Time-Series Decomposition

Query *Python Time Series Decomposition* contains the following script. It makes use of package *statsmodels.api* to calculate the parts of the time series. It is an exact copy of the script in the sister section in Chapter 9, with the difference that in the last lines of the code, the result is not plotted, but rather new columns containing the results are added to data frame *dataset*.

```
# import packages
import pandas as pd
import statsmodels.api as sm
```

The next lines to prepare *dataset* are equal to what we saw in the previous section:

```
# prepare dataset
dataset.Date = pd.to_datetime(dataset.Date)
dataset = dataset.set_index('Date').resample('MS').pad()
y = dataset['Sales Amount'].resample('MS').sum()
```

The model is created via function sm.tsa.seasonal_decompose:

```
# create model
m = sm.tsa.seasonal_decompose(y, model='additive')
```

Columns of the model are then added to the existing *dataset*:

```
# create output
dataset['trend'] = m.trend
dataset['seasonal'] = m.seasonal
dataset['resid'] = m.resid
```

The result in Figure 10-16 shows new columns: *seasonal*, *trend*, and *resid* (residual). Notice that I explicitly changed the data type of the three columns to *Fixed decimal number* (dollar sign left of the column names), as it would be a *Text* otherwise (which is not very useful here). Just click on the "ABC" icon in front of the column name to change the column type.

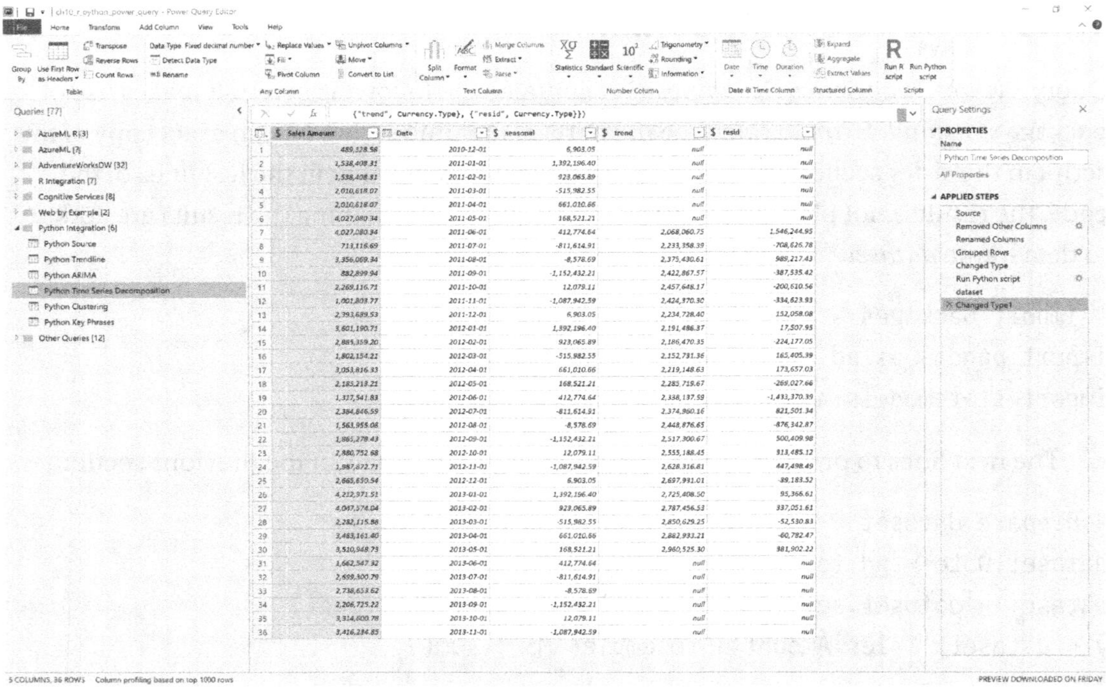

Figure 10-16. *Python script has added three new columns*

Clustering

Clustering is again done through a k-means algorithm (exactly as in Chapter 9). You can see the full script in query *Python Clustering*. Function KMeans from *sklearn* is used as follows:

```
# import packages
from sklearn.cluster import KMeans
```

The model is created over *dataset* and asked to assign three clusters. The result is directly stored in a new column of data frame *dataset*:

```
# create model
dataset["Cluster"] = KMeans(n_clusters=3).fit(dataset).labels_
```

Every existing pair of *OrderQuantity* and *UnitPrice* is assigned to a cluster, shown in Figure 10-17.

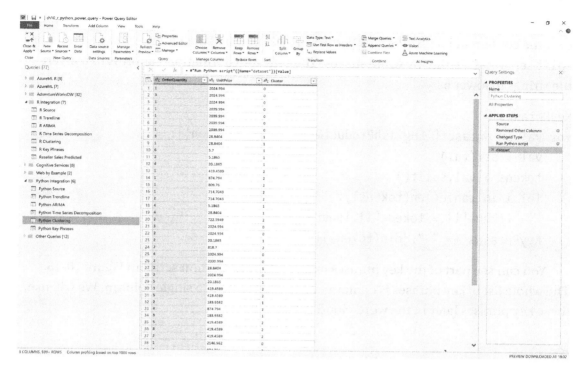

Figure 10-17. *Python script with combinations of OrderQuantity and UnitPrice and their clusters*

Key Phrases

Extracting key phrases is not much different than what we did inside the Python script visual in Chapter 9. Instead of plotting a word cloud, we return a data frame named *output* in this example. (We will plot the key phrases in a later section in this chapter.) First, I load the needed packages:

```
# import packages
import pandas as pd
from wordcloud import STOPWORDS
```

The actual stop words are imported from the *wordcloud* package:

```
# prepare dataset
stopwords = set(STOPWORDS)
```

I then initialize a variable *KeyPhrases* and fill it with the token extracted by looping over the content of *EnglishProductNameAndDescription* of data frame *dataset*. Function split separates a string by white spaces. The only transformation I apply is to convert the strings to lower case.

```
KeyPhrases = ''
for val in dataset['EnglishProductNameAndDescription']:
    val = str(val)
    tokens = val.split()
    for i in range(len(tokens)):
        tokens[i] = tokens[i].lower()
    KeyPhrases += " ".join(tokens)+" "
```

You can see part of the key phrases extracted by the Python script in Figure 10-18. The whole list of key phrases is contained in a single row of a single column. We will use these key phrases later in the word cloud visual.

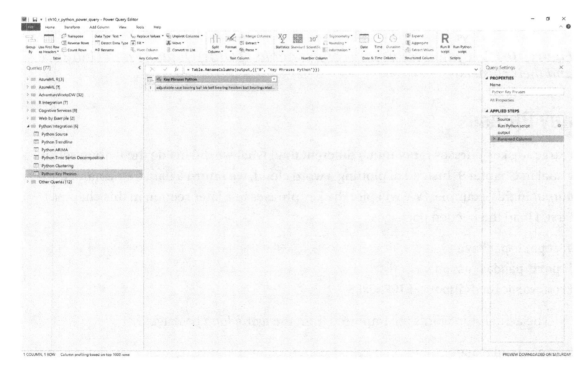

Figure 10-18. *R script's output of key phrases*

Visualize Data Imported and Transformed in R and Python

As soon as data has been loaded into Power BI's data model, its origin does not play any role. You can use any visual to show the data from any table. And "any" really means "any." For instance, you can put the columns from table *R Trendline* into a standard line chart, into an R script visual, and even into a Python script visual. You don't believe me? Then see what I built in Figure 10-19 (report page *Trendline*) for demonstration purposes.

Figure 10-19. *Data transformed with an R script in Power Query shown in a standard visual, R script visual, and Python script visual*

No technical limitation keeps you from even cascading transformations in R and Python in one single Power Query. But probably you want to make a decision on designating either R or Python as the preferred one and would therefore solely use one of the two to transform your data and to visualize your data. Voluntarily limiting the number of different technologies in one single solution is considered a best practice in IT.

There is another important difference, which is explained in the next section.

Trendline

Power Query queries (and their corresponding tables in the Power BI data model) are only updated during a data refresh. The content is *not* influenced by any filter on the report. If you run an R or a Python script visual, then filters are *always* applied on the data exposed to the visual (via data frame *dataset*).

The last two sentences are very important to understand. Their consequence is shown in Figure 10-20 (report page *R Trendline 3*). There, I built a report containing a slicer on the *Date* column and two line charts, each showing *Date, Sales Amount*, and a trendline. (You can find the exact same example for Python in report page *Python Trendline 3* in the example file.)

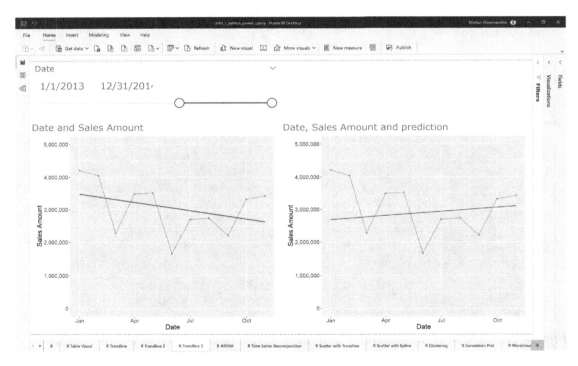

Figure 10-20. *Filters applied on R script visuals with different data sources*

The left line chart is a script visual calculating the trendline as described in Chapter 9 ("Executing R and Python Visualizations"). It calculates the trendline from the data exposed to the visual (only *Date* and *Sales Amount*) before it is plotted together with the data.

The right line chart is a script visual as well. But this script visual is not calculating any trendline; instead, it is simply plotting the given data, which includes *Date* and *Sales Amount* and the data for the trendline (*prediction*).

While the actual data (*Sales Amount*; in green) is identical, the trendline (in blue) in both visuals is totally different. One shows an upward trend, the other a downward trend. The answer to the question, why they are different, lies in the effect of the filter. The left visual is calculating the trendline inside the visual from the data exposed to the visual, which is influenced by the slicer on the *Date* column. A change in the filter not only changes the date range displayed, but also the date range available for the algorithm to calculate the trendline.

A change in the filter does change the date range for the right visual as well, but neither the line for Sales Amount nor the trendline will change its shape. This is because the trendline was pre-calculated in Power Query, and the filter in the report does not influence the behavior of Power Query. Therefore, the trendline is calculated for all available dates, but only a section is displayed.

If this sounds familiar, you are right. Back in section "Trendline in DAX" in Chapter 5 ("Adding Smart Visualizations"), we changed the filters inside the DAX formula. Once we applied it on the actual data available to calculate the trendline (in top section of the formula). In the other example, we applied it on the result of the trendline only (in last lines of the formula). These changes had the same effect as that seen in Figure 10-20.

Neither solution is right or wrong by itself. But tackle the requirements of a report and its filters before you approach one or the other solution.

Time- Series Decomposition

In Figure 10-21, I built a report on different versions of the time-series decomposition. Both visuals on the bottom line are exact copies of the samples demonstrated in Chapter 9 on time-series decomposition. On top, I put the columns from table *R Time Series Decomposition*/*Python Time Series Decomposition* into a stacked area chart. This shows you another example of the difference discussed in the introduction to this section: moving calculations to Power Query frees you up in the decision of what kind of visualization you want to show. With machine learning models applied in Power Query (via R or Python scripts) you can use the output in any visual you like.

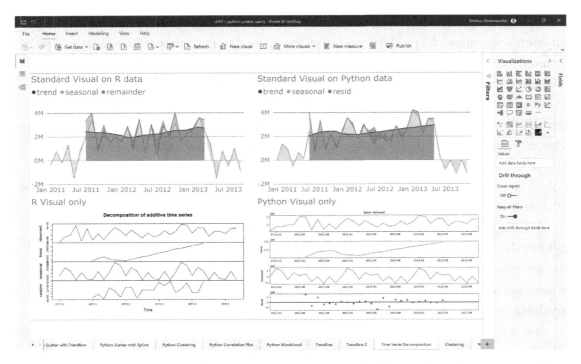

Figure 10-21. *Time-series decomposition in a standard visual and script visuals*

Having two time-series decompositions in the exact same (standard) visual makes it apparent that the algorithms applied in R and Python work slightly differently.

Wordcloud and Key Phrases

For our two word cloud visuals in Chapter 9 ("Executing R and Python Visuals"), we detected the key phrases and plotted the result within the R/Python script visual. This is not so smart. Why? Because, on the one hand, no matter which kind of filter we put on the report, the process of extracting the keywords from the product name and description will always lead to the same result. It is not necessary to extract the key phrases inside the visual. On the other hand, the script in the visual is executed (including the part to extract key phrases) every time the screen refreshes (e.g., when changing a filter or switching to another report and back again).

It would be smarter to decouple the word cloud visual from extracting the key phrases to make the report faster. It would be sufficient to extract the key phrases only when the data model is refreshed. Guess what? R and Python scripts in Power Query only run when the data is refreshed. Therefore, it is wise to put this part of the script from the visual into Power Query from a performance perspective. The difference in performance will depend on the amount of data.

Power BI Desktop comes with a performance analyzer, which you can start in the ribbon via *View ➤ Performance analyzer* (Figure 10-22). There, you can *Start recording*, explicitly *Refresh visual*, *Stop* the refresh when it takes too long, and *Clear* and *Export* the performance analyzer log. The duration is displayed in milliseconds (ms). A value of 1000 represents one second.

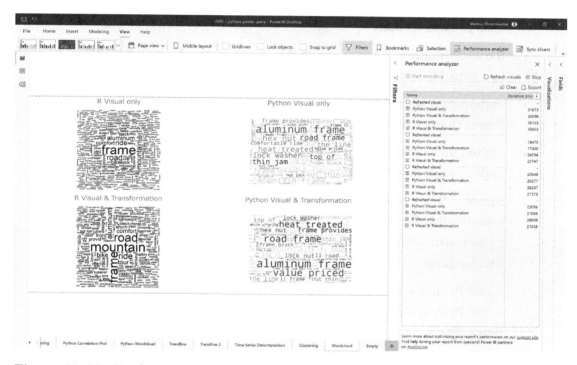

Figure 10-22. *Performance analyzer for visuals of report page word cloud*

From Figure 10-22 we learn that I refreshed the report four times. The duration of the visuals took something between a minimum of 17400 ms and a maximum of 36123 ms. All four executions show the same insights:

- The Python scripts are running constantly faster than the R script visuals on my computer.

- The visuals making use of already existing key phrases in the tables (Python Visual & Transformation / R Visual & Transformation) ran a bit faster than their counterparts that extracted the key phrases inside the script visual.

Decoupling the script had indeed a slight performance effect on the word clouds. Of course, your mileage may vary.

Key Takeaways

R and Python script visuals inside Power Query open up a whole new world, as follows:

- R data source: You can use an R script as the first step in a Power Query to extract data.

- R transformation: You can use an R script as one of the applied steps in a Power Query to filter, transform, and enrich the data.

- Python data source: You can use a Python script as the first step in a Power Query to extract data.

- Python transformation: You can use a Python script as one of the applied steps in a Power Query to filter, transform, and enrich the data.

- Running R or Python in Power Query gives you the freedom to visualize the data in a visualization of your choice. Inside an R or Python visual you are always limited to the visualizations available in that language.

- R and Python scripts in a visual are influenced by filters in the report. R and Python scripts running in Power Query are not influenced by report filters.

- If filtering does not matter, you can move as much of the script into Power Query as possible. This will make the refresh take more time, but the reports will be generated faster. In cases where you refresh the Power BI Desktop file less often than you look at your reports, this is a good idea performance-wise.

Still not enough from R and Python? Learn how to invoke a web service running R and Python scripts in the next chapter.

CHAPTER 11

Execute Machine Learning Models in the Azure Cloud

In this chapter, we are going beyond the boundaries of Power BI Desktop and reaching out to the power of Azure, Microsoft's cloud offering. The easiest way to do this is to use pre-trained models under the name Cognitive Services. *Pre-trained* means that Microsoft's data science team takes care of the model. These models are easy to use because you do not have to train them. If you need more flexibility (as you want to take care of training and tuning the model yourself) or if you want to come up with your very own (or your favorite data scientist's) model then you are more than welcome to do so with the help of Azure Machine Learning Services.

Attention   Everything we have discussed in this book so far was available free of cost. This chapter is an exception. Azure services are priced via subscription models. Some of the functionalities described in this chapter are available in free-trial subscriptions, which allow you to play with the feature. In a production environment, you will most likely hit the limitations of the free subscription. Each section will contain links where you can find out more information about pricing.

© Markus Ehrenmueller-Jensen 2020
M. Ehrenmueller-Jensen, *Self-Service AI with Power BI Desktop*, https://doi.org/10.1007/978-1-4842-6231-3_11

AI Insights

Artificial Intelligence Insights (AI Insights) are a very convenient way of accessing machine learning models. In Power Query (select *Home* ➤ *Transform Data* in Power BI's ribbon to open Power Query) select one of three AI Insights:

- *Text Analytics*

- *Vision*

- *Azure Machine Learning*

Note   Do not confuse the feature explained in this chapter (AI Insights) with the features described in Chapter 2 ("The Insights Feature"): Insights feature in Power BI Desktop and Quick Insights in Power BI Services. All three help you to gain insights but do this in very different ways.

The first two features (*Text Analytics* and *Vision*) require a dedicated Premium capacity with AI workload enabled (`https://docs.microsoft.com/en-us/power-bi/admin/service-premium-what-is`), which includes the resources/costs for using Cognitive Services (no Azure subscription is required). If you receive the following error message (Figure 11-1) or a similar one, you either do not have access to a Premium capacity or the settings are wrong: *No text analytics functions available at this time. Please make sure you have access to a Premium capacity with AI workload enabled. Learn how: `https://aks.ms/enableaiworkload`.* Either buy Premium, correct your settings, or proceed with section "Azure Cognitive Services" to learn how you can access Cognitive Services without Premium capacity.

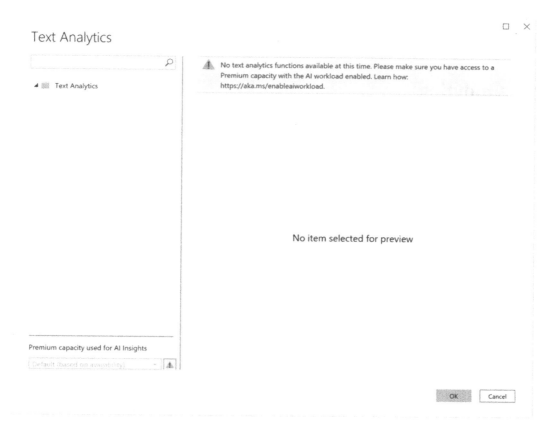

Figure 11-1. AI Insights requires a Premium capacity with AI workload enabled

---

Note   While you are using the graphical user interface to invoke AI Insights, behind the scenes Power BI is automatically creating a Power Query function, which is then used to call the service. In the section about Cognitive Services we will create such a function manually.

The latter (*Azure Machine Learning*) requires read access to an Azure subscription (which contains the trained models).

Azure Machine Learning Services currently exists in three tastes:

- Machine Learning Studio (classic)

- Machine Learning Studio Basic

- Machine Learning Studio Enterprise

We will discuss the integration of the latter two (Machine Learning Studio Basic and Enterprise) in this section and Machine Learning Studio (classic) in a later section.

Text Analytics

Under *Home* ➤ *Text Analytics* (in the *AI Insights* section) we are offered three services:

- Detect language
- Extract key phrases
- Score sentiment

Select *Detect language* and the column for which the language shall be detected (in my case, *EnglishProductNameAndDescription*; see Figure 11-2).

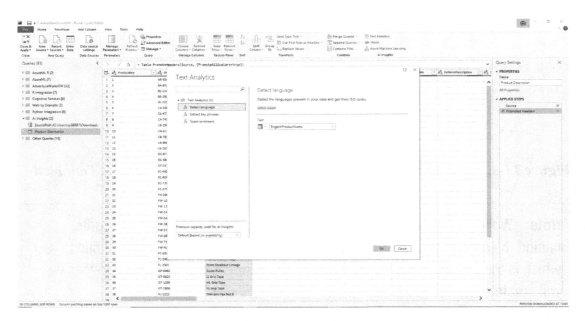

Figure 11-2. *AI Insights Text Analytics: Detect language*

The model returns two columns: *Detected Language Name* (e.g., "English") and *Detected Language ISO Code* (e.g., "en"), as you can see in Figure 11-3.

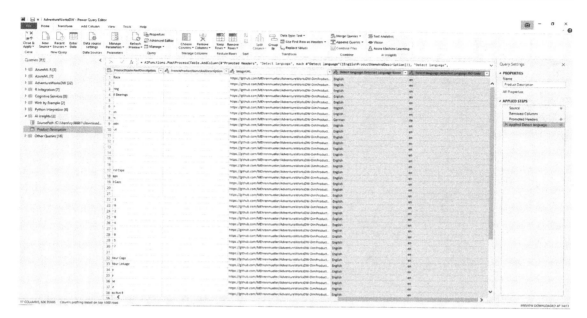

Figure 11-3. *Detected language*

Select *Extract key phrases* as an alternative to what we used in R and Python in the two previous chapters. Pass in the column from which the key phrases shall be extracted as the parameter. If you leave field *Language* empty, *Detect language* is automatically invoked before the key phrases are extracted. As you can see in Figure 11-4, I chose column *EnglishProductNameAndDescription*.

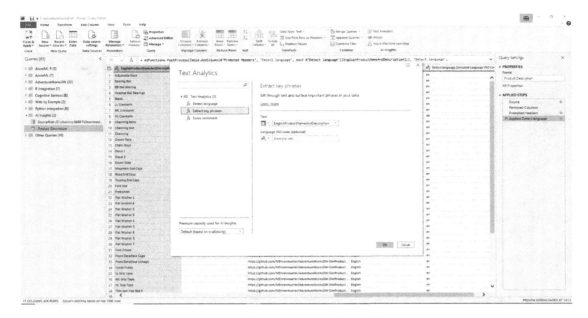

Figure 11-4. *AI Insights Text Analytics: Extract key phrases*

The model returns one column containing all detected key phrases (separated with commas) and a second column containing one row per key phrase. For a word cloud, the first column will be sufficient. Only in special cases will you need one row per key phrase. In Figure 11-5, I scrolled down a bit, as the first rows don't have a description and therefore their key phrases are equal to the product name.

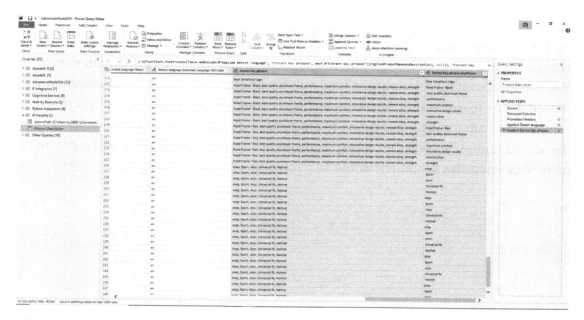

Figure 11-5. *Extracted key phrases*

Select *Score sentiment* and the column for which the sentiment shall be scored (e.g., *EnglishProductNameAndDescription*), as shown in Figure 11-6.

Figure 11-6. *AI Insights Text Analytics: Score sentiment*

One single column, containing the score, is added to the query. The score is a decimal number between zero and one. Zero means that the sentiment of the scored text is negative, one means that the sentiment is positive, and 0.5 points out a neutral text. The products without a description (and only a name) got a broad range of sentiments detected. The reason is that, for example, "flat washer" can be anything in terms of sentiment. If you scroll further down, to products that contain an actual description, the scored sentiment is on the (very) positive side. A product description in the product table with a negative sentiment would be a surprise, I guess.

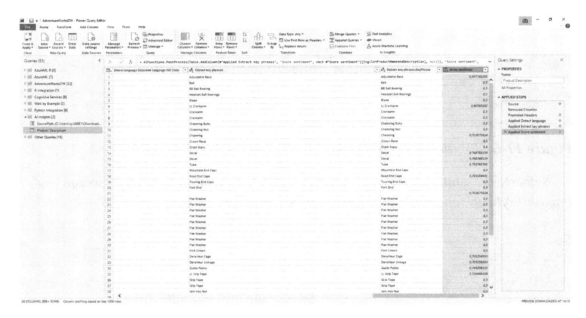

Figure 11-7. *Scored sentiment*

Vision

The *Vision* AI Insight can be applied to a column containing the binary representation of an image or of a URL (web address) of an image file. I chose to not load the images into Power BI because of the size but rather loaded a text column with a link to my (public) GitHub repository, where I keep images for all products of *AdventureWorks*.

Select *Home* ➤ *Vision* (in the *AI Insights* section) and then click on the only option (*Tag images*) and choose the column containing the (link to the) image. In my case, this is column *ImageURL*, as you can see in Figure 11-8.

Figure 11-8. *AI Insights Vision*

This feature returns five values (Figure 11-9):

- *Tags*: a comma-separated list of tags associated with the image

- *Json*: a string in JSON format, containing pairs of tags and confidence
  levels

- *Tag*: the tags are split into a row per tag

- *Confidence*: a decimal between zero and one; the higher the number
  the higher the confidence that the tag truly describes the content of
  the image

- *ErrorMessage*

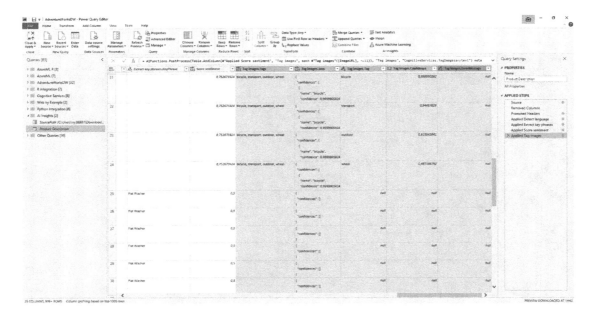

Figure 11-9. *Tags for the ImageURL column*

If you scroll down to the first product that contains a proper image (most products have a dummy image showing text "No Image Available"), which happens to be *Freewheel* (with ProductKey = 21), we get four tags:

- bicycle
- transport
- outdoor
- wheel

The service is very sure that the image shows a *bicycle* (confidence of 1), but less confident that the tag *wheel* fits (confidence of 0.50).

Azure Machine Learning

The basic idea of this feature is that a data scientist can build, tune, train, and maintain a model and publish it as a web service to make it accessible to others in the organization. This web service can then be used, for instance, in the web shop to give a customer recommendations, in a medical application to assist a health check, or—in our case—in Power BI Desktop to enrich the data loaded into the model.

If you want learn more about how to create a model with Azure Machine Learning, please read the information at `https://docs.microsoft.com/en-us/azure/machine-learning/`. I will show you how to build a model with a free subscription of Azure Machine Learning Studio (classic) later in this chapter. A model built with a free subscription can't be used with the AI Insights buttons of Power Query, but I will show you a way around that later in the section "Pre-trained Model in Azure Machine Learning Studio (classic)."

Azure Cognitive Services

Azure Cognitive Services is a whole suite of machine learning services with different application programming interfaces (API):

- Vision

- Speech

- Language

- Search

- Decision

Feel free to explore all the goodness of Cognitive Services here: `https://docs.microsoft.com/en-us/azure/cognitive-services/welcome`. In this section, I picked three Text Analytics services of the Language API for the demos (`https://docs.microsoft.com/en-us/azure/cognitive-services/text-analytics/`) that are analog to the services we saw in the section "AI Insights":

- Language Detection API

- Sentiment API

- Keyphrases API

The first three can easily be accessed if you possess a Premium capacity, as described in the previous section. In this section, I will show you how you can access these services without Premium capacity (or a Pro license, for that matter).

The way to access the services is very similar—the differences are small. I will describe the common part here, and the special parts in the dedicated sections.

1. Acquire a subscription for Azure if you do not possess one yet. It may be a free one as described here: `https://azure.microsoft.com/en-us/free/`.

2. Create Cognitive Services within this subscription via `https://portal.azure.com` for your preferred region. You can find a step-by-step description here: `https://docs.microsoft.com/en-us/azure/cognitive-services/cognitive-services-apis-create-account`.

3. Get Key 1 (or 2) from *Keys and Endpoint* (Figure 11-10).

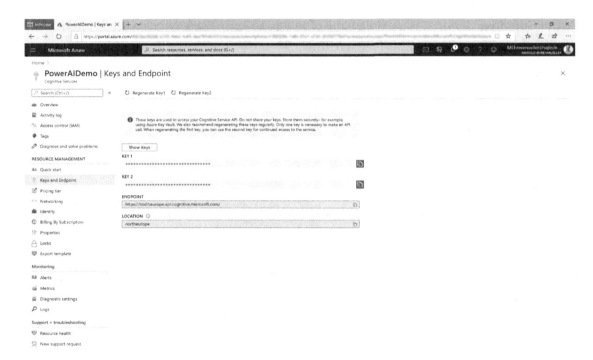

Figure 11-10. *Keys and Endpoint are the key to Cognitive Services*

4. Store the key as a Power Query parameter named *apikey* in the Power BI file. You can change an existing Power Query parameter or create a new one via *Home* ➤ *Manage Parameters* (in the *Parameters* section; see Figure 11-11). We will use this parameter in the scripts later. Having the key defined as a parameter will make it easier to change the key (which you should regularly do as

a security measure to avoid someone's using the service on your
bill; treat the key as securely as you do your passwords; I blurred
my key in the screenshot).

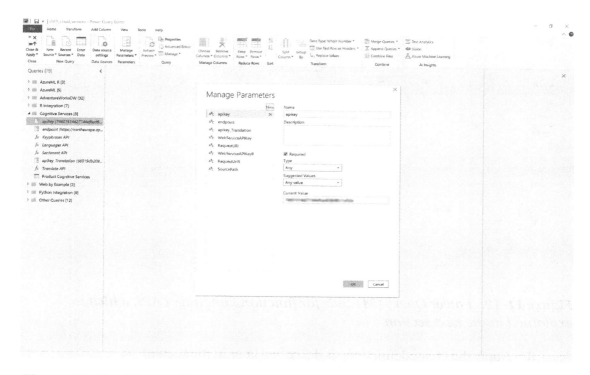

Figure 11-11. *Manage Parameters to change a Power Query parameter or create a new one*

5. Get the endpoint from the same screen. In my case it is `https://northeurope.api.cognitive.microsoft.com`.

6. Add `/text/analytics/v2.0` at the end of the endpoint and store it as a Power Query parameter named *endpoint* in the Power BI file (similar to Power Query parameter *apikey*).

7. Select *New Source* ➤ *Blank query* from Power Query's ribbon. The code I use is based on a script published at `https://docs.microsoft.com/en-us/azure/cognitive-services/text-analytics/tutorials/tutorial-power-bi-key-phrases`. The code and details of the query are described in the specific sections that follow. In Figure 11-12, you can see the definition of the function Languages API.

Figure 11-12. *Power Query (M) code for function Languages API, which is explained in the next section*

8. Apply the created function, as described later in the section "Applying the API Functions."

Language API

Here is the description of the M code to communicate with the Language API.

The code creates a function, which receives one single parameter of type `text`:

```
(text) => let
```

The original script passes a hard-coded string as `apikey`. I changed this to refer to the Power Query parameter created in the previous section:

```
apikey = apikey,
```

The original script passes a hard-coded string as *endpoint*. I changed this to refer to the Power Query parameter created in the previous section, appended with the name of the service (/languages in this case):

```
endpoint = endpoint & "/languages",
```

The next portion is identical to the template from Microsoft's documentation. It converts the text (passed in as a parameter to this function) to JSON, adds structural JSON elements, converts it into a binary, and pastes the *apikey* into the header.

```
jsontext = Text.FromBinary(Json.FromValue(Text.Start(Text.Trim(text),
5000))),
jsonbody = "{ documents: [ { id: ""0"", text: " & jsontext & " } ] }",
bytesbody = Text.ToBinary(jsonbody),
headers = [#"Ocp-Apim-Subscription-Key" = apikey],
```

Then *header* and *text* are passed via Web.Contents to Cognitive Services' *endpoint*:

```
bytesresp = Web.Contents(endpoint, [Headers=headers, Content=bytesbody]),
```

The binary is then converted to a JSON document:

```
jsonresp = Json.Document(bytesresp),
```

And the *detectedLanguages* are returned as the result of this function:

```
returnvalue = jsonresp[documents]{0}[detectedLanguages]
in
returnvalue
```

Don't forget to meaningfully rename the Power Query function. I chose *Languages API*.

Sentiment API

Here is the description of the M code to communicate with the Sentiment API.

The code creates a function, which receives one single parameter of type text:

```
(text) => let
```

The original script passes a hard-coded string as `apikey`. I changed this to refer to the Power Query parameter created in the previous section.

```
apikey = apikey,
```

The original script passes a hard-coded string as *endpoint*. I changed this to refer to the Power Query parameter created in the previous section, appended with the name of the service (`/Sentiment` in this case).

```
endpoint = endpoint & "/Sentiment",
```

The next portion is identical to the template from Microsoft's documentation. It converts the text (passed in as a parameter to this function) to JSON, adds structural JSON elements, converts it into a binary, and passes the *apikey* into the header.

```
jsontext = Text.FromBinary(Json.FromValue(Text.Start(Text.Trim(text),
5000))),
jsonbody = "{ documents: [ { language: ""en"", id: ""0"", text: " &
jsontext & " } ] }",
bytesbody = Text.ToBinary(jsonbody),
headers = [#"Ocp-Apim-Subscription-Key" = apikey],
```

Then *header* and *text* are passed via `Web.Contents` to Cognitive Services' *endpoint*:

```
bytesresp = Web.Contents(endpoint, [Headers=headers, Content=bytesbody]),
```

The binary is then converted to a JSON document:

```
jsonresp = Json.Document(bytesresp),
```

And the result in the form of a *score* is returned from this function:

```
returnvalue = jsonresp[documents]{0}[score]
in
returnvalue
```

Don't forget to meaningfully rename the Power Query function. I chose *Sentiment API*.

Keyphrases API

Here is the description of the M code to communicate with the Keyphrases API.

The code creates a function, which receives one single parameter of type `text`:

```
(text) => let
```

The original script passes a hard-coded string as `apikey`. I changed this to refer to the Power Query parameter created in the previous section.

```
apikey = apikey,
```

The original script passes a hard-coded string as *endpoint*. I changed this to refer to the Power Query parameter created in the previous section, appended with the name of the service (/keyPhrases in this case)

```
endpoint = endpoint & "/keyPhrases",
```

The next portion is identical to the template from Microsoft's documentation. It converts the text (passed in as a parameter to this function) to JSON, adds structural JSON elements, converts it into a binary, and passes the *apikey* into the header.

```
jsontext = Text.FromBinary(Json.FromValue(Text.Start(Text.Trim(text),
5000))),
jsonbody = "{ documents: [ { language: ""en"", id: ""0"", text: " &
jsontext & " } ] }",
bytesbody = Text.ToBinary(jsonbody),
headers = [#"Ocp-Apim-Subscription-Key" = apikey],
```

Then *header* and *text* are passed via `Web.Contents` to Cognitive Services' *endpoint*:

```
bytesresp = Web.Contents(endpoint, [Headers=headers, Content=bytesbody]),
```

The binary is then converted to a JSON document:

```
jsonresp = Json.Document(bytesresp),
```

And the result in the form of *keyPhrases* is returned from this function:

```
returnvalue = Text.Lower(Text.Combine(jsonresp[documents]{0}[keyPhrases], ", "))
in
returnvalue
```

Don't forget to meaningfully rename the Power Query function. I chose *Keyphrases API*.

Applying the API Functions

To apply the functions and get a result from Cognitive Services, select a Power Query and column (I chose column *EnglishProductNameAndDescription* in Power Query *Product Cognitive Services*). Select *Add column* ➤ *Invoke Custom Function* in the ribbon (in the *General* section). Give the new column a meaningful name (e.g., *Language*), and then select the function name (e.g., *Languages API*) and the column to pass to the function (e.g., *EnglishProductNameAndDescription*), as shown in Figure 11-13.

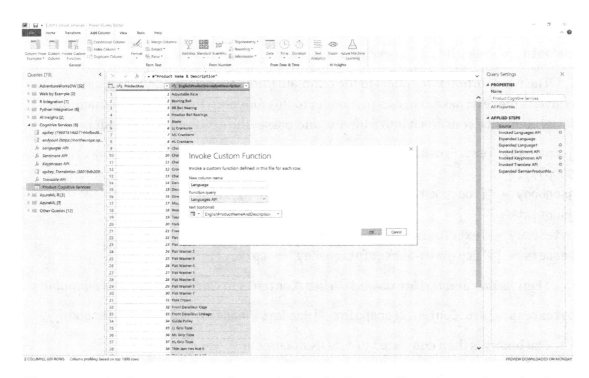

Figure 11-13. *Create a new column via Invoke Custom Function and passing in the function and parameter(s)*

The different functions return the following columns:

- Language API: a list of records which that be expanded (in two steps by clicking the button with the two arrows just right of the column name) to *name* and *iso6391Name* and a *score* (a decimal value between zero and one to represent the confidence that the detected language is the correct one), as you can see in Figure 11-14.

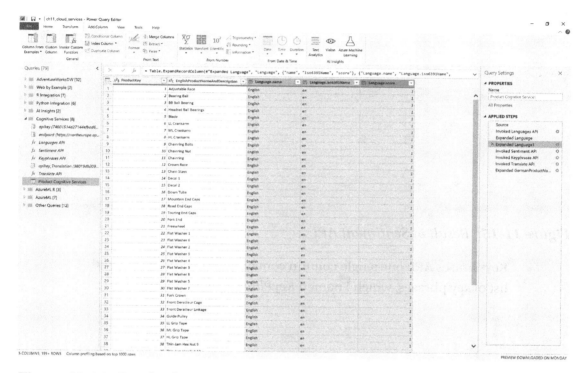

Figure 11-14. *Result of Languages API*

- Sentiment API: a decimal with values between zero and one; zero represents a negative sentiment, one a positive sentiment. I named the result column *Sentiment* (Figure 11-15).

Figure 11-15. *Result of Sentiment API*

- Keyphrases API: one single column containing a comma-separated list of key phrases, which I named *KeyPhrase* (Figure 11-16).

Figure 11-16. *Result of Keyphrases API*

Pre-trained Model in Azure Machine Learning Studio (classic)

Cognitive Services (discussed in the previous section) offers pre-trained models. Why take the effort to use Azure Machine Learning Services (classic) as an extra layer, then? The answer is that this layer offers you certain functionalities to clean and transform the data before and after the pre-trained model is applied. Having these steps within the web service centralizes the necessary logic—no need to apply those steps every time you use the web service in Power Query (or any other application that consumes the web service, for that matter).

To start with Azure Machine Learning Studio (classic), go to `https://studio.azureml.net/` and either sign in with an existing account or sign up with a new account to try it for free (Figure 11-17).

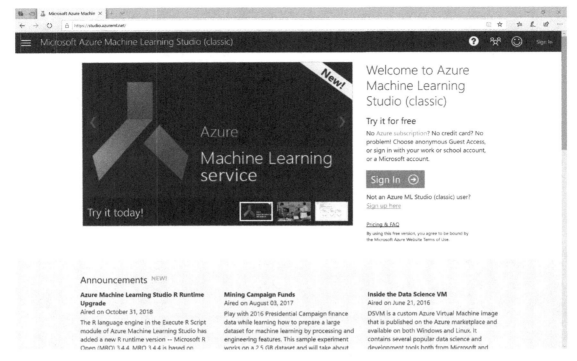

Figure 11-17. *Sign in or sign up to Azure Machine Learning Studio (classic)*

On the left-hand side, you have the choice of *Projects, Experiments, Web Services, Datasets, Trained Models,* and *Settings* (Figure 11-18). We will concentrate on *Experiments* first, and then create a *Web Service* later. Learn more about the other features here: `https://docs.microsoft.com/en-us/azure/machine-learning/studio/what-is-ml-studio`.

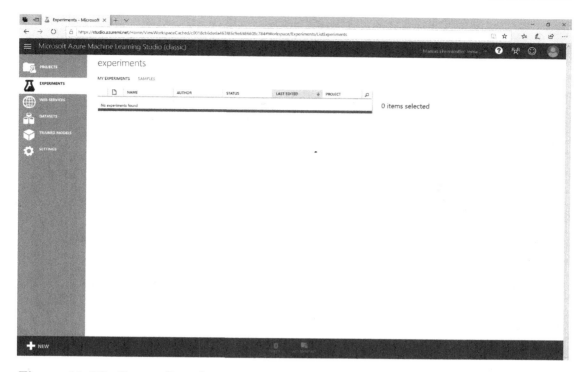

Figure 11-18. *Empty list of experiments in Azure Machine Learning Studio (classic)*

To create a new experiment, click the big plus icon (+ *NEW*) on the bottom left. A long list of ready-to-use examples is offered (Figure 11-19).

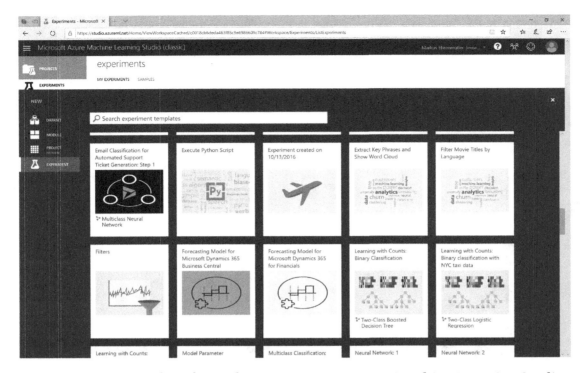

Figure 11-19. *Long list of sample experiments in Azure Machine Learning Studio (classic)*

Find the example *Extract Key Phrases and Show Word Cloud* (as shown in Figure 11-19) by scrolling down or using the search box at the top of the screen, and then click the button *Open in Studio (classic)* (which appears first when you move the mouse cursor over the example).

This experiment contains four elements (Figure 11-20):

- *Book Reviews from Amazon* (as a sample text to have something to play in this experiment)

- *Partition and Sample* (which randomly selects a subset of the data)

- *Extract Key Phrases from Text* (the pre-trained model)

- *Execute R Script* (a script that produces the word cloud)

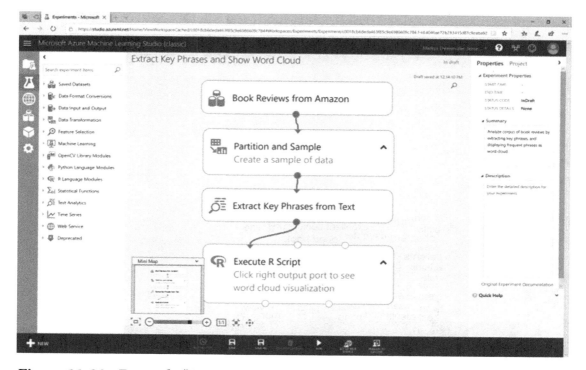

Figure 11-20. *Example "Extract Key Phrases and Show Word Cloud"*

Those four elements are connected by three arrows, which symbolize the flow of the data between the four elements. Click on *Run* at the bottom of the screen to start the experiment. This will take a few seconds. If everything runs successfully, a green checkmark will appear on the elements.

To see the resulting word cloud, right-click on the right output handle (named *R Device (dataset)*) of the *Execute R Script* element and select *Visualize*. The word cloud is listed under *Graphics*. Further information (*Standard Output* and *Standard Error*) is shown as well in Figure 11-21.

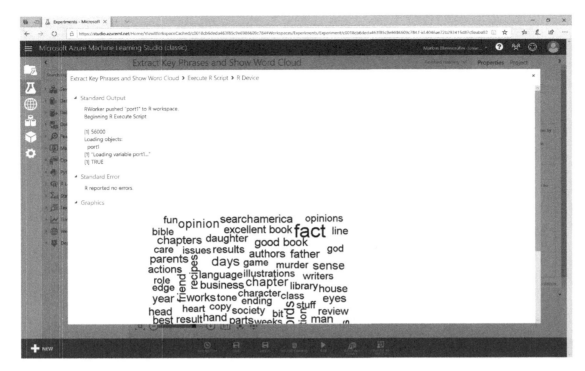

Figure 11-21. *Visualized output*

Visualize is available for the output handles of the other elements as well (after you right-click) and is helpful when you want to debug the experiment (similar to clicking on the *Applied steps* in Power Query).

Before we can convert this experiment into a web service to make it usable in Power Query, we have to apply some modifications:

- Add the item *Edit Metadata* from the list on the left. Select *Data Transformation* ➤ *Manipulation* ➤ *Edit Metadata* (or use the search field to find the item) and drag it below *Book Reviews from Amazon* (Figure 11-22).

- Then, drag the output handle from *Book Reviews from Amazon* to the input handle of *Edit Metadata* (Figure 11-22).

- Select *Edit Metadata* and change its options at the right of the screen. Click on *Launch column selector* and move both *Available Columns* to *Selected Columns* (Figure 11-22) and save the change by clicking on the grey, circled checkmark in the right bottom corner of the flyout window.

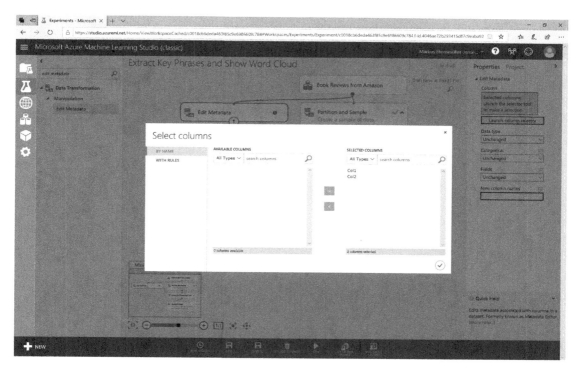

Figure 11-22. *Add Edit Metadata to the experiment, connect it with Book Reviews from Amazon, and move both columns (Col1 and Col2) to Selected Columns*

- Enter `ProductKey, EnglishProductNameAndDescription` into new column names in the properties of *Edit Metadata*, shown on the right (some lines below *Launch column selector*; Figure 11-23).

- Connect the output handle of *Edit Metadata* with the input handle of *Partition and Sample*, as shown in Figure 11-23. (*Book Reviews from Amazon* and *Partition and Sample* are now not connected directly anymore).

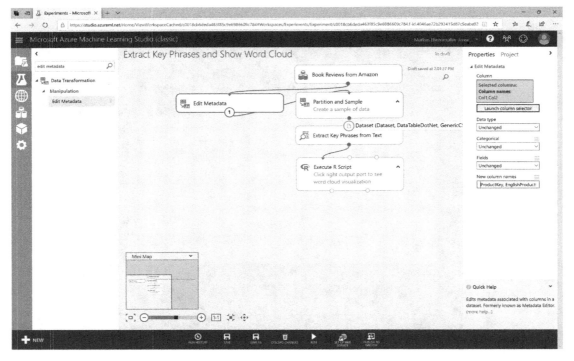

Figure 11-23. *Set new column names and connect Edit Metadata with Partition and Sample*

- Run the experiment to update the metadata of the dataflow (renaming of columns).

- Select *Partition and Sample* and change *Partition or sample mode* to *Head* and enter 700 as the *Number of rows to select*. This filters the input to the first 700 rows (Figure 11-24). This is a safeguard only— feel free to change it to any other number. Keep in mind though, that in the Power BI file provided with this book, we have 606 products for which we want to extract the key phrases.

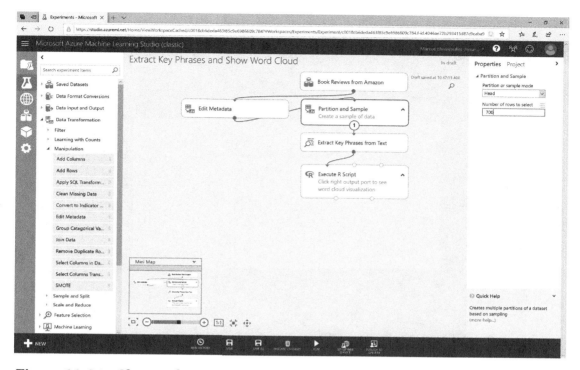

Figure 11-24. *Change the properties of Partition and Sample to Head and 700 rows*

- Select *Extract Key Phrases from Text* and click on *Launch column selector* on the right. Remove *Col2* and select *EnglishProductNameAndDescription* instead (Figure 11-25).

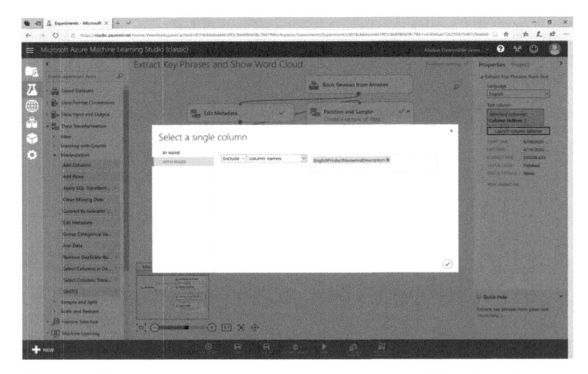

Figure 11-25. *Launch column selector to change the name of the column in Extract Key Phrases from Text*

- Select a new item *Add columns* (*Data Transformation* ➤ *Manipulation*) and drag it below *Extract Key Phrases from Text* (Figure 11-26). Connect the output handle of *Partition and Sample* with the left input handle of *Add columns*, and the output handle of *Extract Key Phrases from Text* with the right input handle of *Add columns*. This allows us to return both the input columns and the key phrases. There are no properties to set.

- Right-click *Execute R Script* and select *Delete* as we do not need the word cloud visualization (Figure 11-26).

- Rename the experiment to *Extract Key Phrases* (as we do not create a word cloud anymore). Just click on the title at the top of the screen and delete the second part of the title (Figure 11-26).

- Rearrange the items so that no item overlaps a data-flow arrow. See my version in Figure 11-26.

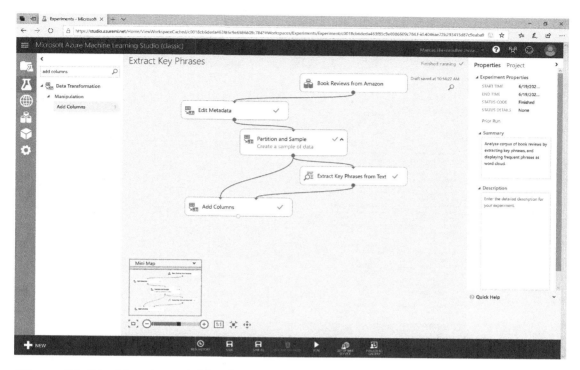

Figure 11-26. *New item Add columns added and connected, Execute R Script removed, and all items rearranged to please the eye*

- Run the experiment. If there is an error (red *x*-icon at an item) carefully read the error message and double-check if you executed the steps exactly as described.

- Right-click the output handle of *Add columns* and *Visualize* the result. You should see the rows for three columns (*ProductKey*, *EnglishProductNameAndDescription*, and *Key Phrases*), as in Figure 11-27.

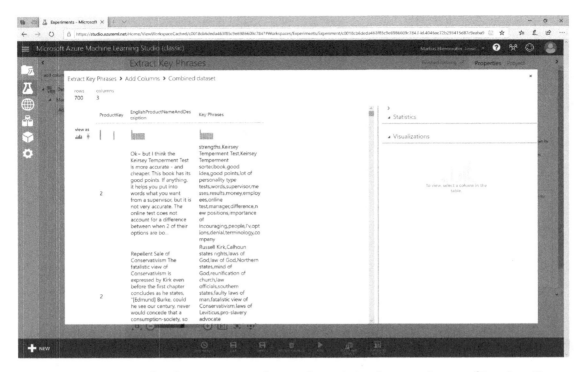

Figure 11-27. *Result of experiment shows the original two columns (ProductKey and EnglishProductNameAndDescription) and the generated Key Phrases*

We can consider this experiment as successful and can convert it to a web service. (Don't be afraid, the web service is created without our having to write a single line of code.) Click *Set Up a Web Service* on the bottom (right of *Run*). An invisible hand then adds two items: *Web service input* and *Web service output*. Read the explanation carefully and click through the four steps of the wizard by selecting *Next* (Figure 11-28).

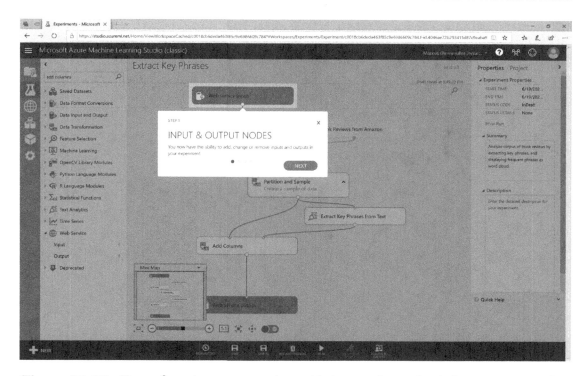

Figure 11-28. *Transforming an experiment into a web service is just a matter of mouse clicks*

After the *Web service input* and *Web service output* are added we must run the experiment again before we can click on *Deploy web service*. Keep the shown API key as a secret—it will allow anybody to use the web service on your dime (or make your free trial run out sooner than you expected). Therefore, I blurred my key in Figure 11-29.

Figure 11-29. *The API key is necessary to connect to the web service*

Copy the displayed API key and return to Power Query and create a new Power Query parameter (*Home* ➤ *Manage Parameters* ➤ *New Parameter*) with name *WebServiceAPIKey*, as demonstrated in Figure 11-30. (If you are working with the demo file you don't need to create a new parameter, but you need your API key, as the one in the file is void.) Paste the API key as the *Current Value* for the parameter.

Figure 11-30. *Power Query parameter WebServiceAPIKey holds the API key*

Return to the web browser and click on *Request/Response* (Figure 11-31). Copy the displayed Request URI and create another Power Query parameter with name *RequestURI*, analogous to *WebServiceAPIKey* in the previous paragraph. (If you are working with the demo file you don't need to create a new parameter). Paste the *RequestURI* as the *Current Value* for the parameter.

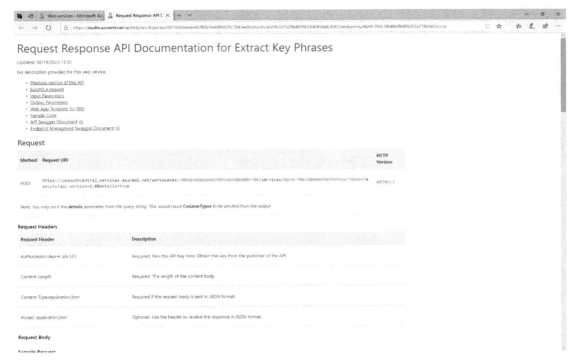

Figure 11-31. *Request URI of the newly created web service*

The idea of both parameters is to have the connection info to the web service easily at hand in case you need it in different places in Power Query and in case the information changes. You should handle the API key and Request URI like a password: don't disclose them to third parties and change them regularly. To change the API key you simply delete the web service and then create it from the experiment again.

Next, let's actually invoke the brand-new web service we just created. I use code developed by Gerhard Brückl, which he blogged about at `https://blog.gbrueckl.at/2016/06/score-powerbi-datasets-dynamically-azure-ml/`. The concept comes with three Power Query functions:

- `ToAzureMLJson` converts the data into JSON (similar to what we did earlier in section "Azure Cognitive Services").

- `AzureMLJsonToTable` converts the data from JSON into a Power Query table (similar to what we did earlier in section "Azure Cognitive Services").

- CallAzureMLService uses the two preceding functions and the Power Query parameters (*apikey* and *RequestURI*) to send the content of a whole table to the Azure Machine Learning Services web service.

Here is the M code for function ToAzureMLJson. In Power Query select *Home* ➤ *New Source* ➤ *Blank Query* and paste the whole code that follows to create the function. The function has one mandatory parameter (the *input* that should be converted) and one optional parameter (*inputName*). Depending on the data type, different transformations are applied to generate valid JSON code. Here is the code:

```
// Source: https://blog.gbrueckl.at/2016/06/score-powerbi-datasets-
dynamically-azure-ml/
let
    ToAzureMLJson= (input as any, optional inputName as text) as text =>
let
    inputNameFinal = if inputName = null then "input" else inputName,
    transformationList = {
        [Type = type time, Transformation = (value_in as time) as text =>
        """" & Time.ToText(value_in, "hh:mm:ss.sss") & """"],
        [Type = type date, Transformation = (value_in as date) as text =>
        """" & Date.ToText(value_in, "yyyy-MM-dd") & """"],
        [Type = type datetime, Transformation = (value_in as datetime) as
        text => """" & DateTime.ToText(value_in, "yyyy-MM-ddThh:mm:ss.sss"
        & """")],
        [Type = type datetimezone, Transformation = (value_in as
        datetimezone) as text => """" & DateTimeZone.ToText(value_in,
        "yyyy-MM-ddThh:mm:ss.sss") & """"],
        [Type = type duration, Transformation = (value_in as duration) as
        text => ToAzureMLJson(Duration.TotalSeconds(value_in))],
        [Type = type number, Transformation = (value_in as number) as text
        => Number.ToText(value_in, "G", "en-US")],
        [Type = type logical, Transformation = (value_in as logical) as
        text => Logical.ToText(value_in)],
        [Type = type text, Transformation = (value_in as text) as text =>
        """" & value_in & """"],
```

```
[Type = type record, Transformation = (value_in as record)
as text =>
                    let
                        GetFields = Record.FieldNames(value_in),
                        FieldsAsTable = Table.FromList(GetFields,
                        Splitter.SplitByNothing(), {"FieldName"},
                        null, ExtraValues.Error),
                        AddFieldValue = Table.AddColumn(
                        FieldsAsTable, "FieldValue", each
                        Record.Field(value_in, [FieldName])),
                        AddJson = Table.AddColumn(AddFieldValue,
                        "__JSON", each
                        ToAzureMLJson([FieldValue])),

                        jsonOutput = "[" & Text.Combine(AddJson
                        [__JSON], ",") & "]"
                    in
                        jsonOutput
                    ],
    [Type = type table, Transformation = (value_in as table) as text =>
                    let
                        BufferedInput = Table.Buffer(value_in),
                        GetColumnNames = Table.ColumnNames(
                        BufferedInput),
                        ColumnNamesAsTable = Table.FromList(
                        GetColumnNames , Splitter.SplitByNothing(),
                        {"FieldName"}, null, ExtraValues.Error),
                        ColumnNamesJson = """ColumnNames"": ["""
                        & Text.Combine(ColumnNamesAsTable[Field
                        Name], """, """) & """]",

                        AddJson = Table.AddColumn(value_in,
                        "__JSON", each ToAzureMLJson(_)),
                        ValuesJson = """Values"": [" & Text.
                        Combine(AddJson[__JSON], ",#(lf)") & "]",
```

```
                        jsonOutput = "{""Inputs""": { "&
                        inputNameFinal & ": {" & ColumnNamesJson &
                        "," & ValuesJson & "} },
                        ""GlobalParameters""": {} }"
                in
                        jsonOutput
                ],
        [Type = type list, Transformation = (value_in as list) as text =>
        ToAzureMLJson(Table.FromList(value_in, Splitter.SplitByNothing(),
        {"ListValue"}, null, ExtraValues.Error))],
        [Type = type binary, Transformation = (value_in as binary) as text
        => """0x" & Binary.ToText(value_in, 1) & """"],
        [Type = type any, Transformation = (value_in as any) as text => if
        value_in = null then "null" else """" & value_in & """"]
    },
    transformation = List.First(List.Select(transformationList , each
    Value.Is(input, _[Type]) or _[Type] = type any))[Transformation],
    result = transformation(input)
in
    result
in
    ToAzureMLJson
```

Function AzureMLJsonToTable does the opposite of function ToAzureMLJson: it converts the JSON back to a format Power Query can understand. In Power Query, select *Home* ➤ *New Source* ➤ *Blank Query* and paste the whole code that follows to create the function:

```
// https://blog.gbrueckl.at/2016/06/score-powerbi-datasets-dynamically-
azure-ml/
let
    AzureMLJsonToTable = (azureMLResponse as binary) as any =>
let

    WebResponseJson = Json.Document(azureMLResponse ,1252),
    Results = WebResponseJson[Results],
    output1 = Results[Output],
```

```
    value = output1[value],
    BufferedValues = Table.Buffer(Table.FromRows(value[Values])),
    ColumnNameTable = Table.AddIndexColumn(Table.FromList(value[ColumnNames],
    Splitter.SplitByNothing(), {"NewColumnName"}, null, ExtraValues.Error),
    "Index", 0, 1),
    ColumnNameTable_Values = Table.AddIndexColumn(Table.FromList(Table.
    ColumnNames(BufferedValues), null, {"ColumnName"}), "Index", 0, 1),

    RenameList = Table.ToRows(Table.RemoveColumns(Table.Join(
    ColumnNameTable_Values, "Index", ColumnNameTable, "Index"),{"Index"})),
    RenamedValues = Table.RenameColumns(BufferedValues, RenameList),

    ColumnTypeTextTable = Table.AddIndexColumn(Table.FromList(
    value[ColumnTypes], Splitter.SplitByNothing(), {"NewColumnType_Text"},
    null, ExtraValues.Error), "Index", 0, 1),
    ColumnTypeText2Table = Table.AddColumn(ColumnTypeTextTable,
    "NewColumnType", each
      if Text.Contains([NewColumnType_Text], "Int") then type number
else if Text.Contains([NewColumnType_Text], "DateTime") then type datetime
else if [NewColumnType_Text] = "String" then type text
else if [NewColumnType_Text] = "Boolean" then type logical
else if [NewColumnType_Text] = "Double" or [NewColumnType_Text] = "Single"
then type number
else if [NewColumnType_Text] = "datetime" then type datetime
else if [NewColumnType_Text] = "DateTimeOffset" then type datetimezone
else type any),
    ColumnTypeTable  = Table.RemoveColumns(ColumnTypeText2Table
    ,{"NewColumnType_Text"}),

    DatatypeList = Table.ToRows(Table.RemoveColumns(Table.Join(
    ColumnNameTable, "Index", ColumnTypeTable, "Index"),{"Index"})),
    RetypedValues = Table.TransformColumnTypes(RenamedValues, DatatypeList,
    "en-US"),
```

```
    output = RetypedValues
in

    output

in

    AzureMLJsonToTable
```

The preceding two functions are helper functions. Next comes the function we will apply on the data (as we applied the functions in section "Azure Cognitive Services"). The function has three mandatory parameters: *RequestURI* (which we fill with the Power Query parameter of the same name), *WebServiceAPIKey* (which we fill with Power Query parameter *apikey*), and *TableToScore* is the name of the Power Query query we want to use as the input of the web service. Make sure that the columns in the query match the type and name of the web service. The fourth parameter, *Timeout*, is optional. In Power Query, select *Home ➤ New Source ➤ Blank Query* and paste the whole code that follows to create the function:

```
// https://blog.gbrueckl.at/2016/06/score-powerbi-datasets-dynamically-
azure-ml/
let
    AzureMLJsonToTable = (
        RequestURI as text,
        WebServiceAPIKey as text,
        TableToScore as table,
        optional Timeout as number
    ) as any =>
let
    WebTimeout = if Timeout = null then #duration(0,0,0,100) else
    #duration(0,0,0,Timeout) ,

    WebServiceContent = ToAzureMLJson(TableToScore),

    RequestURI1 = RequestURI,
    WebServiceAPIKey1 = WebServiceAPIKey,

    WebResponse = Web.Contents(RequestURI,
        [Content = Text.ToBinary(WebServiceContent),
         Headers = [Authorization="Bearer " & WebServiceAPIKey,
```

```
                    #"Content-Type"="application/json",
                    Accept="application/json"],
        Timeout = WebTimeout]),

    output = AzureMLJsonToTable(WebResponse)
in
    output

in

    AzureMLJsonToTable
```

Finally, we can put the pieces together and apply the function. I created a Power Query named *Product Name & Description* (which I disabled to not actually load any table with that name into Power BI) and referenced it in the following Power Query (again, select *Home* ➤ *New Source* ➤ *Blank Query*); then, I pasted in the following code.

First, reference the existing query, *Product Name & Description*:

```
let
    TableToScore = #"Product Name & Description",
```

Then, call function `CallAzureMLService` and pass in Power Query parameters *RequestURI* and *WebServiceAPIKey* as the first parameter and then the previously defined *TableToScore*. As we are OK with the default timeout, the fourth parameter is `null`:

```
Source = CallAzureMLService(RequestURI, WebServiceAPIKey, TableToScore,
null),
```

All columns are returned as strings; therefore, change the type of *ProductKey* to `Int64.Type` (whole number) and return all rows:

```
#"Changed Type" = Table.TransformColumnTypes(Source,{{"ProductKey",
Int64.Type}})
in
#"Changed Type"
```

Your Own Model in Azure Machine Learning Studio (classic)

To keep it simple, we will duplicate the Azure Machine Learning Studio (classic) experiment from the previous example (*Extract Key Phrases*) and exchange the pretrained component *Extract Key Phrases from Text* with a script, written in R, which we have already seen in Chapters 9 and 10. I will guide you through the necessary steps:

- Open Azure Machine Learning Studio (classic) via `https://studio.azureml.net/` and select *Experiments* and the experiment created in the previous section (which I named *Extract Key Phrases*). On the bottom, click *Save as* and enter the new name (I chose *Extract Key Phrases R*; shown in Figure 11-31) and click the circled checkmark.

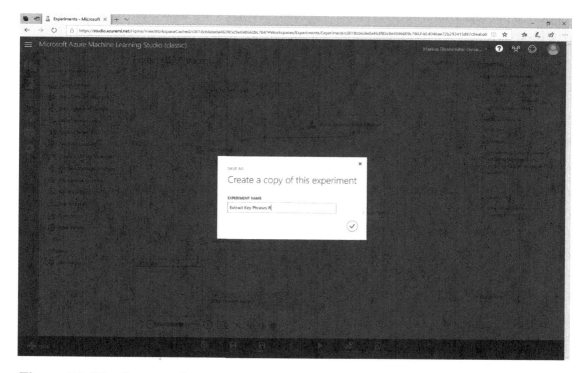

Figure 11-32. *Open and save an experiment under a different name to copy it*

- Remove item *Extract Key Phrases from Text*.

- Select a new item *Execute R Script* (*R Language Modules*) and drag it to the position where *Extract Key Phrases from Text* previously was.

335

- Connect the output of *Partition and Sample* with the first input of *Execute R Script.*

- Connect the first output of *Execute R Script* with the second input of *Add Columns.*

- Copy and paste the following R script into field *R Script* (Figure 11-33).

Unlike Power Query, Azure Machine Learning Studio (classic) does not automatically inject data as a data frame. We must call `maml.mapInputPort` with the *portnumber* the previous item is connected to.

```
# Map 1-based optional input ports to variables
dataset <- maml.mapInputPort(1) # class: data.frame
```

Then, we load package `tm`, which contains algorithms for text mining. Find a list of supported packages here: `https://docs.microsoft.com/en-us/azure/machine-learning/studio-module-reference/r-packages-supported-by-azure-machine-learning`.

```
# load packages
library(tm)
```

Next, we run the same script as in Chapter 10, when we extracted key phrases to show the word cloud:

```
# prepare data
vc <- VCorpus(VectorSource(dataset$EnglishProductNameAndDescription))
vc <- tm_map(vc, content_transformer(tolower))
vc <- tm_map(vc, removeNumbers)
vc <- tm_map(vc, removeWords, stopwords())
vc <- tm_map(vc, removePunctuation)
vc <- tm_map(vc, stripWhitespace)
vc <- tm_map(vc, stemDocument)
```

Finally, we need to return the result as a data frame and pass it to `maml.mapOutputPort` (Figure 11-33):

```
# Select data.frame to be sent to the output Dataset port
df <- data.frame(text = sapply(corpus_clean, paste, collapse = " "),
stringsAsFactors = FALSE)
maml.mapOutputPort("df")
```

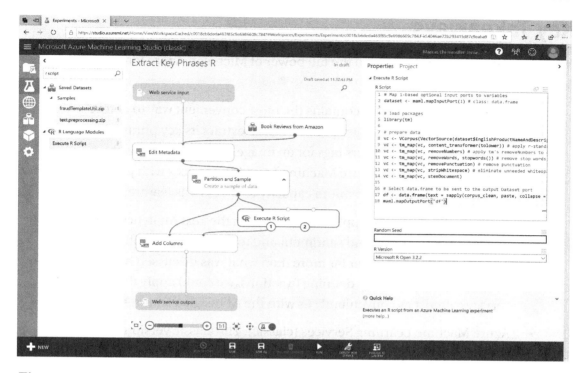

Figure 11-33. *Extract Key Phrases from Text is replaced with Execute R Script*

If you prefer Python instead, you can exchange the *Execute R Script* with *Execute Python Script* and insert the Python script from Chapter 10.

Follow the same steps as in the previous section to create a web service and call it from Power Query:

- Run the experiment.

- Deploy web service.

- Copy and paste the API key and Request URI into Power Query parameters.

- Apply function `CallAzureMLService` on a text column.

Key Takeaways

Reaching out to the cloud gives you the full power of Microsoft Azure for Power BI in the following ways:

- AI Insights: This section contains the most convenient way to detect the language of a text or its sentiment or to extract its key phrases. The Vision service delivers tags for an image. You can access web services published in Azure Machine Learning Services with just a few clicks. You need a Premium capacity to access this feature.

- Cognitive Services: We applied services from the Text Analytics API to detect the language and sentiment and to extract the key phrases. The offered services cover far more than what was discussed here and are ever growing. By defining functions, we could apply the services almost as conveniently as with the AI Insights features.

- Azure Machine Learning Services (classic): The classic version of Azure Machine Learning Services is available as a free trial. This service allows us to add logic before and after we call either a pre-trained model or run a script in R or Python. By defining functions, we could apply the services almost as conveniently as with the AI Insights features.

Believe it or not, this was the very last chapter. Thanks for taking time to read until the final page. I hope you learned something from this book. So long, and thanks for all the fish!

Index

Printed in the United States
By Bookmasters